Mental Aerobics

Math Puzzles for Everyone

Volume One

James Albert Moehlenbrock

Mind Booster Books
Travelers Rest, South Carolina

Mental Aerobics
Math Puzzles for Everyone by James Albert Moehlenbrock
Volume One

Published by Mind Booster Books
21 Newport Drive, Suite 104
Travelers Rest, South Carolina 29690 USA

Printed and Bound in the United States of America by Create Space, an Amazon Company

Cover design by Create Space

ISBN-13: 978-0-9815537-1-9
ISBN-10: 0-9815537-1-0

Library of Congress Control Number: 2008901479

Dedication Page

This book is dedicated to the kids at the orphanage, Chiang Mai Home for Boys in northern Thailand, whose keen interest in solving these math puzzles helped inspire me to produce this book of puzzles so that others may enjoy them and benefit from them, too.

However, special recognition is earned by one particular nine-year-old boy at the orphanage, A-Jewel Merlare, who is actually a mountain tribe kid from the Lahu Tribe. A-Jewel complied with all the suggestions in this book. He memorized the multiplication tables and achieved 100% accuracy on all four timed tests. He solved every one of the puzzles in this book and was able to solve some of the advanced ones with seven and eight unknowns in less than one minute. On many occasions he correctly worked the long ones in his head before writing the complete answer down. Truthfully, the book has more puzzles than I had intended it to have simply because A-Jewel would daily ask for more challenging puzzles which I created for him and then would add to the book.

A-Jewel loved the challenge of math so much that his curiosity for learning seemed unquenchable. Consequently, he learned to work problems involving fractions and decimals using all four operations and he could work with percentages and could do so without hesitation. Even as a nine-year-old he had started learning to solve Algebraic equations. Because I worked with the entire age range of students at the orphanage, I actually believe A-Jewel knew more math than any of the high school students.

Since A-Jewel so willingly exemplified what the reader is asked to do in this book, and because he proved what mastering the multiplication tables can do for a person in other areas of math, along with the surprise results of enhancing his self-confidence and awakening a joy he now has for exploring new math concepts, I wholeheartedly add A-Jewel to the dedication page of this book, *Mental Aerobics -- Math Puzzles for Everyone*. Furthermore, I want to personally thank A-Jewel for his excellent work habits and enthusiastic attitude and for truly being a joy to work with.

--- James Albert Moehlenbrock

About the Author

James Albert Moehlenbrock is a semi-retired, internationally experienced high school counselor and mathematics teacher having taught math for more than 30 years in USA, Germany, Vietnam (in the late 1990's), and in Thailand. He has traveled extensively into 56 countries stretched out over five continents and has lived over one-third of his adult life outside the USA. He always has been fascinated with solving puzzles and has introduced his own puzzles to nearly every age group across many different cultures. Consequently, he has learned that nearly everyone enjoys the challenge of solving a puzzle.

Jim holds master's degrees in counseling and in mathematics. He grew up in the northern portion of South Carolina, and just recently has returned to his South Carolina home having completed a wonderful thirteen-year experience teaching mathematics in Southeast Asia.

Acknowledgments

When I retired from South Carolina Public Schools as a high school math teacher and school counselor, I moved to Thailand to do volunteer work. I taught math for four years in a mountain tribe school and during the evenings, I worked with an orphanage. Later, I decided to work only with the orphanage trying to help the kids improve their math skills. In the section where I worked, there were 160 kids from age six to nineteen.

My first acknowledgment goes to these kids, as they are the ones who truly endorsed my puzzles. Actually, they had very little interest in the academics, especially math. They would much rather play football (soccer) or one of their many other Asian sports. However, presenting them with math in the form of a puzzle sparked interest. They liked them. In fact, they enjoyed the puzzles, the kind you have in this book, so much that they asked for more and more puzzles. They wanted me to give the same puzzle to several kids at once to see who could find a valid solution first. Indeed, those kids amazed me in how fast they could come up with a solution that worked. In order for them to get an edge on the others, I believe they studied the tables on their own so they could find puzzle solutions even more quickly.

As you read on the Dedication Page, nine-year old A-Jewel worked every one of the puzzles in this book. In the Answer Section, occasionally, I showed one of A-Jewel's solutions with different puzzles to share his simple but practical approach to solving a puzzle – starting out with testing "ones" and going from there with larger numbers. As much as A-Jewel and I enjoyed playing badminton together, he would often choose puzzle solving to badminton (but that might have been due to how hot it was outside).

So I want to thank the orphans at Chiang Mai Home for Boys for showing me how much they truly enjoyed my math puzzles and how those puzzles triggered in them a new fondness for math. Yes, math can really be fun as their enthusiasm clearly showed me. Of course, I knew math was fun already.

My second acknowledgment and thank you go to Bob and Jane MacCallum, retired teachers themselves, who also liked my math puzzles. They are the ones who initially encouraged me to write this book. So thanks, Bob and Jane, for that needed push.

None of this work would have been possible had it not been for the Director of Chiang Mai Home for Boys, Mrs.Boonmee Savangtume, giving me permission to work at the orphanage. Not only was she supportive of the work I was doing, but each year she provided me with the letter I needed for applying for a long-term-stay Thai Visa.

It was when I started with this book that I realized working these puzzles can truly sharpen one's own mind, even a senior citizen's mind which is what I've got, as I could see an improved difference in my own thinking speed.

I wish you the same benefit.

TABLE OF CONTENTS

Mental Aerobics

Math Puzzles for Everyone

Volume One

The purpose of *Mental Aerobics – Math Puzzles for Everyone* is to have fun and to sharpen one's thinking. Every individual puzzle is based on the multiplication tables (or multiplication facts, as they are sometimes called) from one through nine. Therefore, the puzzles are actually suitable for anyone from the third grader, to whom the tables are generally introduced, through adulthood. If you are a senior citizen, these are puzzles you and your grandchildren can work on together or in competition with one another. Everyone should benefit. The kids will improve their basic math skills while the older folks will be exercising their minds and possibly warding off Alzheimer's. But these puzzles are not just for senior citizens and grandkids. They are for everyone in-between, as well.

You are right, zero is a part of the tables we learned, but zero has been left out of these puzzles, because zero can negate a lot of a puzzle defeating the purpose of improving our skills with the rest of the tables.

Now, if you can't recall all the multiplication facts, there is a simple exercise in the Help Section toward the back of the book that can help make you an expert very quickly. Also, there are four timed tests in the back that can tell you how proficient you already are. It might just be a good idea for you to read the Help Section beginning on page 89 before continuing, as you may find those pages beneficial.

Let's assume you've studied the Help Section and are now ready to begin attacking the puzzles. We'll begin with two examples so you'll be sure you understand.

Example One:

$(5 \times \square) + 1 = \square \times 4$ If you replace the first box with 3 and the second box with 4, you'll have a solution: $(5 \times 3) + 1 = 4 \times 4$.

$$15 + 1 = 16$$
$$16 = 16$$

With some equations, you'll be asked to find only one solution. However, with other equations, you may be asked to find more than one solution. As you can see with the equation you just had, 3 and 4 work as one solution, but so do 7 and 9.

Let's look at our second example – a more challenging one.

Example Two:

$(5 \times \square) + (\square \times 6) + 1 = \square \times 8$

With a challenger like this one, you might be asked to find nine different solutions. Can you figure this one out? You may not, just yet, but when you work up to this level, your thinking will be much sharper than it is now, and you'll surprise yourself with how quickly you'll be finding solutions, and that will be fun!

Here are nine solutions:

5, 5, and 7
9, 3, and 8
3, 8, and 8
1, 7, and 6
7, 6, and 9
3, 4, and 5
1, 3, and 3
5, 1, and 4
7, 2, and 6

As a reminder, you can only use a single digit from one through nine in a box. Also, to give you a little encouragement, in order to work cross-word puzzles, which are verbal mind boosters, you need to work with a pool of thousands of different vocabulary words. Whereas, working with these math puzzles, a non-verbal activity, you only need to know 45 different math facts, if you count 3 X 4 as being the same fact as 4 X 3. What an easier pool to work with!

Why don't you go ahead and begin, and then if you decide you need more help, you may choose to review the helpful tips in the Help Section, after all. Now, have some fun with the one thousand, three hundred twenty-two puzzles ahead of you.

Chapter One   One-Solution Puzzles

In this group of puzzles, you are only looking for one solution. Later, in other groups, you'll be working with puzzles where you'll be asked to find more than twenty different solutions. There is a Solution Section in the back of the book where you may check for answers. Some of you may find these puzzles too easy, and other of you may find them to be about right as starters. They do gradually get more difficult as you turn the pages.

1. $(9 \text{ X } \square) - 1 = \square \text{ X } 4$

2. $(9 \text{ X } \square) + 1 = \square \text{ X } 5$

3. $(9 \text{ X } \square) + 1 = \square \text{ X } 2$

4. $(9 \text{ X } \square) + 1 = \square \text{ X } 4$

5. $(9 \text{ X } \square) - 1 = \square \text{ X } 7$

6. $(9 \text{ X } \square) - 1 = (\square \text{ X } 5) + 1$

7. $(9 \text{ X } \square) + 1 = (\square \text{ X } 7) - 1$

8. $(9 \text{ X } \square) + 1 = \square \text{ X } 7$

9. $(9 \text{ X } \square) - 1 = \square \text{ X } 5$

10. $(9 \text{ X } \square) + 1 = (\square \text{ X } 8) - 1$

11. $(9 \text{ X } \square) + 1 = (\square \text{ X } 5) - 1$

12. $(9 \text{ X } \square) - 1 = (\square \text{ X } 8) + 1$

13. $(9 \text{ X } \square) + 1 = \square \text{ X } 8$

14. $(9 \text{ X } \square) + 1 = (\square \text{ X } 4) - 1$

15. $(9 \text{ X } \square) - 1 = (\square \text{ X } 4) + 1$

16. $(9 \text{ X } \square) - 1 = (\square \text{ X } 2) + 1$

17. $(8 \text{ X } \square) - 1 = (\square \text{ X } 3) + 1$

18. $(8 \text{ X } \square) + 1 = \square \text{ X } 3$

19. $(8 \text{ X } \square) + 1 = (\square \text{ X } 5) - 1$

20. $(8 \text{ X } \square) - 1 = (\square \text{ X } 7) + 1$

21. $(8 \text{ X } \square) - 1 = \square \text{ X } 5$

22. $(8 \text{ X } \square) + 1 = (\square \text{ X } 9) - 1$

23. $(8 \text{ X } \square) + 1 = \square \text{ X } 5$

24. $(8 \text{ X } \square) - 1 = (\square \text{ X } 5) + 1$

25. $(8 \text{ X } \square) - 1 = \square \text{ X } 3$

26. $(8 \text{ X } \square) + 1 = (\square \text{ X } 7) - 1$

27. $(8 \text{ X } \square) + 1 = \square \text{ X } 7$

28. $(8 \text{ X } \square) + 1 = (\square \text{ X } 3) - 1$

29. $(8 \text{ X } \square) - 1 = (\square \text{ X } 9) + 1$

30. $(8 \text{ X } \square) - 1 = \square \text{ X } 9$

31. $(7 \text{ X } \square) - 1 = (\square \text{ X } 4) + 1$

32. $(7 \text{ X } \square) + 1 = \square \text{ X } 3$

33. $(7 \text{ X } \square) - 1 = \square \text{ X } 5$

34. $(7 \text{ X } \square) - 1 = (\square \text{ X } 3) + 1$

35. $(7 \text{ X } \square) - 1 = \square \text{ X } 9$

36. $(7 \text{ X } \square) + 1 = (\square \text{ X } 6) - 1$

37. $(7 \times \square) - 1 = (\square \times 2) + 1$

38. $(7 \times \square) + 1 = (\square \times 2) - 1$

39. $(7 \times \square) - 1 = \square \times 4$

40. $(7 \times \square) + 1 = \square \times 5$

41. $(7 \times \square) + 1 = (\square \times 5) - 1$

42. $(7 \times \square) + 1 = \square \times 6$

43. $(7 \times \square) + 1 = (\square \times 4) - 1$

44. $(7 \times \square) + 1 = \square \times 9$

45. $(7 \times \square) - 1 = (\square \times 8) + 1$

46. $(7 \times \square) + 1 = (\square \times 8) - 1$

47. $(7 \times \square) - 1 = \square \times 8$

48. $(7 \times \square) - 1 = (\square \times 9) + 1$

49. $(7 \times \square) + 1 = (\square \times 3) - 1$

50. $(6 \times \square) + 1 = (\square \times 5) - 1$

51. $(6 \times \square) + 1 = \square \times 5$

52. $(6 \times \square) - 1 = (\square \times 7) + 1$

53. $(6 \times \square) - 1 = \square \times 7$

54. $(5 \times \square) - 1 = \square \times 9$

55. $(5 \times \square) - 1 = \square \times 7$

56. $(5 \times \square) + 1 = \square \times 7$

57. $(5 \times \square) + 1 = \square \times 8$

58. $(5 \times \square) - 1 = (\square \times 9) + 1$

59. $(5 \times \square) - 1 = \square \times 8$

60. $(5 \times \square) - 1 = (\square \times 7) + 1$

61. $(5 \times \square) - 1 = (\square \times 6) + 1$

62. $(5 \times \square) - 1 = \square \times 6$

63. $(5 \times \square) + 1 = \square \times 9$

64. $(5 \times \square) + 1 = (\square \times 9) - 1$

65. $(5 \times \square) + 1 = (\square \times 8) - 1$

66. $(5 \times \square) - 1 = \square \times 3$

67. $(4 \times \square) + 1 = (\square \times 9) - 1$

68. $(4 \times \square) + 1 = \square \times 7$

69. $(4 \times \square) - 1 = \square \times 9$

70. $(4 \times \square) - 1 = (\square \times 7) + 1$

71. $(4 \times \square) + 1 = (\square \times 7) - 1$

72. $(4 \times \square) - 1 = (\square \times 9) + 1$

73. $(3 \times \square) + 1 = (\square \times 7) - 1$

74. $(3 \times \square) - 1 = \square \times 8$

75. $(3 \times \square) - 1 = \square \times 7$

76. $(3 \times \square) - 1 = (\square \times 8) + 1$

77. $(3 \times \square) + 1 = \square \times 8$

78. $(2 \times \square) + 1 = (\square \times 7) - 1$

79. $(2 \times \square) + 1 = \square \times 9$

80. $(2 \times \square) - 1 = (\square \times 5) + 1$

81. $(2 \times \square) + 1 = (\square \times 9) - 1$

82. $(3 \times \square) + (\square \times 6) + 1 = (\square \times 2) - 1$

Indeed, they are getting more difficult, but your thinking skills are getting sharper, too.

83. (3 X □) + (□ X 7) + 1 = □ X 2   84. (3 X □) + (□ X 8) + 1 = (□ X 2) – 1

85. (4 X □) + (□ X 9) = □ X 3   86. (4 X □) + (□ X 8) + 1 = (□ X 3) – 1

87. (4 X □) + (□ X 7) = □ X 2   88. (4 X □) + (□ X 7) – 1 = (□ X 2) + 1

89. (5 X □) + (□ X 9) = □ X 3   90. (5 X □) + (□ X 6) – 1 = (□ X 2) + 1

91. (5 X □) + (□ X 6) = □ X 2   92. (5 X □) + (□ X 6) + 1 = (□ X 2) – 1

93. (5 X □) + (□ X 7) + 1 = □ X 2   94. (5 X □) + (□ X 8) – 1 = (□ X 2) + 1

95. (5 X □) + (□ X 8) = □ X 2

96. (5 X □) + (□ X 9) – 1 = □ X 2

97. (5 X □) + (□ X 9) + 1 = (□ X 3) – 1

98. (6 X □) + (□ X 9) – 1 = (□ X 4) + 1

99. (6 X □) + (□ X 7) = □ X 3

100. (6 X □) + (□ X 9) + 1 = (□ X 4) – 1

101. (6 X □) + (□ X 7) - 1 = (□ X 2) + 1

102. (6 X □) + (□ X 8) + 1 = (□ X 3) – 1

103. (7 X □) + (□ X 8) = □ X 4

104. (7 X □) + (□ X 8) – 1 = □ X 3

105. (7 X □) + (□ X 9) – 1 = (□ X 3) + 1

106. (7 X □) + (□ X 8) + 1 = (□ X 3) – 1

107. (7 X □) + (□ X 9) + 1 = □ X 4

108. (7 X □) + (□ X 9) + 1 = □ X 3

109. (7 X □) + (□ X 8) + 1 = □ X 3

110. $(8 \times \square) + (\square \times 9) = \square \times 6$

111. $(8 \times \square) + (\square \times 9) - 1 = \square \times 3$

112. $(8 \times \square) + (\square \times 9) + 1 = (\square \times 3) - 1$

113. $(8 \times \square) + (\square \times 9) + 1 = \square \times 4$

You have just completed all of the one-solution problems. In the next chapter you'll be finding two different solutions for each puzzle.

Chapter Two   **Two-Solution Puzzles**

114. $(3 \text{ X } \square) + (\square \text{ X } 8) - 1 = (\square \text{ X } 2) + 1$

115. $(3 \text{ X } \square) + (\square \text{ X } 7) + 1 = (\square \text{ X } 2) - 1$

116. $(4 \text{ X } \square) + (\square \text{ X } 8) - 1 = (\square \text{ X } 5) + 1$

117. $(4 \text{ X } \square) + (\square \text{ X } 6) = \square \text{ X } 3$

118. $(8 \text{ X } \square) - 1 = (\square \text{ X } 2) + 1$

119. $(8 \text{ X } \square) + 1 = (\square \text{ X } 2) - 1$

120. $(8 \text{ X } \square) + 1 = (\square \text{ X } 6) - 1$

121. $(8 \text{ X } \square) = \square \text{ X } 6$

122. $(7 \text{ X } \square) - 1 = \square \text{ X } 3$

123. $(7 \text{ X } \square) - 1 = (\square \text{ X } 6) + 1$

124. $(7 \text{ X } \square) + 1 = \square \text{ X } 4$

125. $(6 \text{ X } \square) + 1 = (\square \text{ X } 7) - 1$

126. $(6 \text{ X } \square) - 1 = (\square \text{ X } 5) + 1$

127. $(6 \text{ X } \square) - 1 = (\square \text{ X } 8) + 1$

128. $6 \text{ X } \square = \square \text{ X } 8$

129. $(5 \text{ X } \square) + 1 = \square \text{ X } 3$

130. $(5 \text{ X } \square) + 1 = (\square \text{ X } 6) - 1$

131. $(5 \text{ X } \square) - 1 = (\square \text{ X } 4) + 1$

132. $(5 \text{ X } \square) + 1 = \square \text{ X } 4$

133. $(5 \text{ X } \square) + 1 = (\square \text{ X } 4) - 1$

134. $(5 \text{ X } \square) - 1 = (\square \text{ X } 2) + 1$

135. $(5 \times \square) + 1 = \square \times 2$

136. $(5 \times \square) - 1 = (\square \times 3) + 1$

137. $(5 \times \square) + 1 = (\square \times 3) - 1$

138. $(4 \times \square) + 1 = \square \times 3$

139. $(4 \times \square) - 1 = (\square \times 5) + 1$

140. $(4 \times \square) - 1 = (\square \times 3) + 1$

141. $(4 \times \square) - 1 = \square \times 5$

142. $(4 \times \square) + 1 = (\square \times 3) - 1$

143. $4 \times \square = \square \times 3$

144. $(4 \times \square) + 1 = (\square \times 5) - 1$

145. $(3 \times \square) - 1 = \square \times 5$

146. $(3 \times \square) + 1 = \square \times 5$

147. $(3 \times \square) + 1 = (\square \times 4) - 1$

148. $(3 \times \square) - 1 = (\square \times 5) + 1$

149. $(3 \times \square) - 1 = \square \times 4$

150. $(3 \times \square) + 1 = (\square \times 2) - 1$

151. $(3 \times \square) - 1 = (\square \times 4) + 1$

152. $(3 \times \square) + 1 = \square \times 7$

153. $(2 \times \square) - 1 = (\square \times 8) + 1$

154. $(2 \times \square) - 1 = \square \times 5$

155. $(2 \times \square) + 1 = (\square \times 8) - 1$

156. $(2 \times \square) + 1 = (\square \times 5) - 1$

157. $(2 \times \square) - 1 = (\square \times 6) + 1$

158. $(2 \times \square) + 1 = \square \times 5$

159. $(4 \times \square) + (\square \times 8) + 1 = \square \times 3$

160. $(4 \times \square) + (\square \times 5) - 1 = (\square \times 2) + 1$

161. $(4 \times \square) + (\square \times 5) = \square \times 2$

162. $(4 \times \square) + (\square \times 9) + 1 = \square \times 3$

163. $(4 \times \square) + (\square \times 8) - 1 = (\square \times 3) + 1$

164. $(5 \times \square) + (\square \times 8) = \square \times 4$

165. $(5 \times \square) + (\square \times 6) = \square \times 3$

166. $(5 \times \square) + (\square \times 7) - 1 = \square \times 3$

167. $(5 \times \square) + (\square \times 8) + 1 = \square \times 4$

168. $(5 \times \square) + (\square \times 7) + 1 = (\square \times 3) - 1$

169. $(5 \times \square) + (\square \times 8) + 1 = \square \times 3$

170. $(5 \times \square) + (\square \times 6) + 1 = (\square \times 3) - 1$

171. $(5 \times \square) + (\square \times 7) - 1 = \square \times 2$

172. $(5 \times \square) + (\square \times 7) + 1 = \square \times 3$

173. $(5 \times \square) + (\square \times 8) = \square \times 3$

174. $(5 \times \square) + (\square \times 9) - 1 = \square \times 3$

175. $(5 \times \square) + (\square \times 9) + 1 = \square \times 4$

176. $(6 \times \square) + (\square \times 9) + 1 = (\square \times 5) - 1$

177. $(6 \times \square) + (\square \times 7) = \square \times 4$

178. $(6 \times \square) + (\square \times 9) - 1 = \square \times 5$

179. $(6 \times \square) + (\square \times 9) = \square \times 4$

180. $(6 \times \square) + (\square \times 7) - 1 = (\square \times 3) + 1$

181. $(6 \times \square) + (\square \times 7) + 1 = \square \times 3$

182. $(6 \times \square) + (\square \times 7) + 1 = (\square \times 4) - 1$

183. $(6 \times \square) + (\square \times 8) - 1 = \square \times 3$

184. $(7 \times \square) + (\square \times 9) - 1 = (\square \times 5) + 1$

185. $(7 \times \square) + (\square \times 9) - 1 = \square \times 4$

186. $(7 \times \square) + (\square \times 8) - 1 = \square \times 4$

187. $(7 \times \square) + (\square \times 8) - 1 = (\square \times 3) + 1$

188. $(7 \times \square) + (\square \times 9) = \square \times 6$

189. $(7 \times \square) + (\square \times 9) - 1 = (\square \times 4) + 1$

190. $(7 \times \square) + (\square \times 9) = \square \times 5$

191. $(7 \times \square) + (\square \times 9) + 1 = (\square \times 4) - 1$

192. $(7 \times \square) + (\square \times 8) + 1 = (\square \times 4) - 1$

193. $(7 \times \square) + (\square \times 9) + 1 = (\square \times 5) - 1$

194. $(7 \times \square) + (\square \times 9) - 1 = (\square \times 4) + 1$

195. $(7 \times \square) + (\square \times 9) + 1 = (\square \times 3) - 1$

196. $(8 \times \square) + (\square \times 9) - 1 = \square \times 5$

197. $(8 \times \square) + (\square \times 9) - 1 = (\square \times 5) + 1$

198. $(8 \times \square) + (\square \times 9) + 1 = (\square \times 5) - 1$

199. $(8 \times \square) + (\square \times 9) + 1 = (\square \times 6) - 1$

200. $(8 \times \square) + (\square \times 9) - 1 = (\square \times 4) + 1$

201. $(8 \times \square) + (\square \times 9) + 1 = \square \times 5$

202. $(8 \times \square) + (\square \times 9) + 1 = (\square \times 4) - 1$

203. $(8 \times \square) + (\square \times 9) = \square \times 5$

204. $(8 \times \square) + (\square \times 9) + 1 = \square \times 3$

You've just finished the two-solution puzzles. Now with this next section, try to figure out three solutions to each problem. Solving these puzzles may even be getting easier for you, as your confidence grows. Have some more fun!

Chapter Three   **Three-Solution Puzzles**

205. $(9 \times \square) = \square \times 6$

206. $(6 \times \square) - 1 = (\square \times 2) + 1$

207. $6 \times \square = \square \times 4$

208. $(4 \times \square) - 1 = (\square \times 6) + 1$

209. $(3 \times \square) + 1 = \square \times 2$

210. $3 \times \square = \square \times 2$

211. $(3 \times \square) - 1 = (\square \times 2) + 1$

212. $(2 \times \square) - 1 = \square \times 3$

213. $(2 \times \square) + 1 = (\square \times 6) - 1$

214. $(2 \times \square) + 1 = (\square \times 3) - 1$

215. $(3 \times \square) + (\square \times 4) + 1 = (\square \times 2) - 1$

216. $(3 \times \square) + (\square \times 5) + 1 = \square \times 2$

217. $(3 \times \square) + (\square \times 6) - 1 = (\square \times 2) + 1$

218. $(3 \times \square) + (\square \times 7) - 1 = \square \times 2$

219. $(4 \times \square) + (\square \times 9) - 1 = \square \times 3$

220. $(4 \times \square) + (\square \times 7) - 1 = \square \times 3$

221. $(4 \times \square) + (\square \times 7) = \square \times 3$

222. $(4 \times \square) + (\square \times 9) - 1 = (\square \times 3) + 1$

223. $(4 \times \square) + (\square \times 6) + 1 = (\square \times 3) - 1$

224. $(4 \times \square) + (\square \times 8) + 1 = (\square \times 5) - 1$

225. $(5 \times \square) + (\square \times 6) + 1 = (\square \times 4) - 1$

226. (5 X □) + (□ X 7) + 1 = (□ X 4) − 1

227. (5 X □) + (□ X 7) + 1 = □ X 4

228. (5 X □) + (□ X 9) − 1 = (□ X 4) + 1

229. (5 X □) + (□ X 9) − 1 = □ X 4

230. (5 X □) + (□ X 7) − 1 = (□ X 3) + 1

231. (5 X □) + (□ X 8) − 1 = (□ X 3) + 1

232. (5 X □) + (□ X 9) = □ X 4

233. (5 X □) + (□ X 6) − 1 = □ X 3

234. (5 X □) + (□ X 8) + 1 = (□ X 4) − 1

235. (5 X □) + (□ X 9) − 1 = (□ X 3) + 1

236. (6 X □) + (□ X 9) + 1 = (□ X 8) − 1

237. (6 X □) + (□ X 8) + 1 = (□ X 5) − 1

238. (6 X □) + (□ X 9) + 1 = □ X 7

239. (6 X □) + (□ X 9) + 1 = □ X 5

240. (6 X □) + (□ X 7) − 1 = (□ X 4) + 1

241. (6 X □) + (□ X 7) + 1 = □ X 4

242. (6 X □) + (□ X 8) = □ X 5

243. (6 X □) + (□ X 8) − 1 = (□ X 5) + 1

244. (6 X □) + (□ X 9) − 1 = (□ X 5) + 1

245. (6 X □) + (□ X 9) − 1 = □ X 4

246. (6 X □) + (□ X 8) − 1 = (□ X 3) + 1

247. (6 X □) + (□ X 7) + 1 = (□ X 3) − 1

248. $(7 \times \square) + (\square \times 8) - 1 = (\square \times 6) + 1$

249. $(7 \times \square) + (\square \times 8) + 1 = (\square \times 5) - 1$

250. $(7 \times \square) + (\square \times 9) + 1 = \square \times 6$

251. $(7 \times \square) + (\square \times 9) - 1 = (\square \times 6) + 1$

252. $(7 \times \square) + (\square \times 9) + 1 = \square \times 5$

253. $(7 \times \square) + (\square \times 8) + 1 = (\square \times 6) - 1$

254. $(7 \times \square) + (\square \times 8) + 1 = \square \times 5$

255. $(7 \times \square) + (\square \times 8) - 1 = \square \times 5$

256. $(7 \times \square) + (\square \times 9) - 1 = \square \times 5$

257. $(7 \times \square) + (\square \times 8) - 1 = (\square \times 5) + 1$

258. $(7 \times \square) + (\square \times 8) - 1 = (\square \times 4) + 1$

259. $(8 \times \square) + (\square \times 9) - 1 = \square \times 7$

260. $(8 \times \square) + (\square \times 9) - 1 = \square \times 6$

261. $(8 \times \square) + (\square \times 9) - 1 = (\square \times 6) + 1$

262. $(8 \times \square) + (\square \times 9) + 1 = \square \times 7$

263. $(8 \times \square) + (\square \times 9) = \square \times 7$

264. $(8 \times \square) + (\square \times 9) + 1 = (\square \times 7) - 1$

That's it, folks, for the three-solution puzzles. Now you get to graduate to the big 4.

Chapter Four   Four-Solution Puzzles

These problems have four different solutions. See if you can find them all. Even if you just find one solution, you are doing well. But with finding all four, you are doing outstandingly well. Good luck!

265. $(2 \times \square) + (\square \times 8) + 1 = (\square \times 3) - 1$

266. $(3 \times \square) + (\square \times 8) = \square \times 4$

267. $(3 \times \square) + (\square \times 9) + 1 = (\square \times 5) - 1$

268. $(2 \times \square) + (\square \times 9) = \square \times 3$

269. $(2 \times \square) + (\square \times 9) + 1 = (\square \times 4) - 1$

270. $(2 \times \square) + (\square \times 9) + 1 = (\square \times 3) - 1$

271. $(6 \times \square) + 1 = (\square \times 4) - 1$

272. $(4 \times \square) + 1 = (\square \times 2) - 1$

Notice the minus sign between the first two groups in the next two problems.

273. $(8 \times \square) - (\square \times 7) - 1 = \square \times 9$

274. $(8 \times \square) - (\square \times 9) = \square \times 7$

Now, we are back to the plus sign.

275. $(3 \times \square) + (\square \times 4) = \square \times 2$

276. $(3 \times \square) + (\square \times 5) - 1 = \square \times 2$

277. $(3 \times \square) + (\square \times 9) - 1 = (\square \times 4) + 1$

278. $(4 \times \square) + (\square \times 9) + 1 = (\square \times 5) - 1$

279. $(4 \times \square) + (\square \times 9) = \square \times 6$

280. $(4 \times \square) + (\square \times 5) + 1 = \square \times 3$

281. $(4 \times \square) + (\square \times 5) + 1 = (\square \times 3) - 1$

282. $(4 \times \square) + (\square \times 6) - 1 = (\square \times 3) + 1$

283. $(4 \times \square) + (\square \times 6) + 1 = \square \times 3$

284. $(4 \times \square) + (\square \times 7) - 1 = (\square \times 3) + 1$

285. $(4 \times \square) + (\square \times 8) - 1 = \square \times 3$

286. $(4 \times \square) + (\square \times 9) + 1 = \square \times 5$

287. $(5 \times \square) + (\square \times 9) = \square \times 6$

288. $(5 \times \square) + (\square \times 6) = \square \times 4$

289. $(5 \times \square) + (\square \times 8) + 1 = (\square \times 6) - 1$

290. $(5 \times \square) + (\square \times 8) - 1 = (\square \times 4) + 1$

291. $(5 \times \square) + (\square \times 9) + 1 = (\square \times 6) - 1$

292. $(5 \times \square) + (\square \times 6) - 1 = (\square \times 4) + 1$

293. $(5 \times \square) + (\square \times 7) - 1 = (\square \times 4) + 1$

294. $(5 \times \square) + (\square \times 7) - 1 = \square \times 4$

295. $(6 \times \square) + (\square \times 9) - 1 = (\square \times 7) + 1$

296. $(6 \times \square) + (\square \times 9) + 1 = (\square \times 7) - 1$

297. $(6 \times \square) + (\square \times 7) + 1 = \square \times 5$

298. $(6 \times \square) + (\square \times 7) - 1 = (\square \times 5) + 1$

299. $(6 \times \square) + (\square \times 8) + 1 = (\square \times 7) - 1$

300. $(6 \times \square) + (\square \times 8) - 1 = \square \times 5$

301. $(6 \times \square) + (\square \times 7) = \square \times 5$

302. $(6 \times \square) + (\square \times 7) - 1 = \square \times 5$

303. $(6 \times \square) + (\square \times 8) = \square \times 4$

304. $(6 \times \square) + (\square \times 8) + 1 = \square \times 5$

305. $(6 \times \square) + (\square \times 7) + 1 = (\square \times 5) - 1$

306. $(7 \times \square) + (\square \times 8) - 1 = \square \times 6$

307. $(7 \times \square) + (\square \times 9) + 1 = \square \times 8$

308. $(7 \times \square) + (\square \times 9) + 1 = (\square \times 8) - 1$

309. $(7 \times \square) + (\square \times 9) - 1 = \square \times 6$

310. $(7 \times \square) + (\square \times 9) + 1 = (\square \times 6) - 1$

311. $(7 \times \square) + (\square \times 8) + 1 = \square \times 6$

312. $(8 \times \square) + (\square \times 9) - 1 = (\square \times 7) + 1$

313. $(8 \times \square) + (\square \times 9) + 1 = \square \times 6$

That's it for the fours. Now we're ready for the next group. You are right – the fives! You answered that so quickly. It just goes to show how sharp you've already become.

Chapter Five   **Five-Solution Puzzles**

Enjoy finding five solutions for each of the following puzzles.

314. $(2 \times \square) + (\square \times 9) = \square \times 4$

315. $(2 \times \square) + (\square \times 9) + 1 = \square \times 4$

316. $(2 \times \square) + (\square \times 9) - 1 = (\square \times 4) + 1$

317. $(2 \times \square) + (\square \times 9) - 1 = \square \times 3$

318. $(2 \times \square) + (\square \times 8) = \square \times 3$

In the next several problems, notice that the sign between the first two groups has changed to minus. However, you are still finding five solutions.

319. $(9 \times \square) - (\square \times 6) = \square \times 8$

320. $(9 \times \square) - (\square \times 8) = \square \times 6$

321. $(8 \times \square) - (\square \times 4) = \square \times 9$

322. $(8 \times \square) - (\square \times 7) = \square \times 9$

323. $(8 \times \square) - (\square \times 7) + 1 = \square \times 9$

324. $(8 \times \square) - (\square \times 9) = \square \times 4$

Now we are back to using the plus sign again.

325. $(2 \times \square) + (\square \times 5) - 1 = (\square \times 2) + 1$

326. $(3 \times \square) + (\square \times 9) - 1 = \square \times 5$

327. $(3 \times \square) + (\square \times 8) = \square \times 6$

328. (3 X □) + (□ X 9) – 1 = □ X 4

329. (3 X □) + (□ X 8) – 1 = □ X 4

330. (3 X □) + (□ X 9) + 1 = □ X 7

331. (3 X □) + (□ X 9) + 1 = □ X 5

332. (3 X □) + (□ X 8) + 1 = (□ X 4) – 1

333. (3 X □) + (□ X 5) – 1 = (□ X 2) + 1

334. (3 X □) + (□ X 4) – 1 = (□ X 2) + 1

335. (4 X □) + (□ X 9) = □ X 8

336. (4 X □) + (□ X 5) – 1 = □ X 3

337. (4 X □) + (□ X 8) + 1 = (□ X 7) – 1

338. (4 X □) + (□ X 8) – 1 = □ X 5

339. (4 X □) + (□ X 9) – 1 = (□ X 6) + 1

340. (4 X □) + (□ X 9) = □ X 5

341. (4 X □) + (□ X 9) – 1 = □ X 5

342. (4 X □) + (□ X 9) – 1 = (□ X 5) + 1

343. (4 X □) + (□ X 9) + 1 = (□ X 6) – 1

344. (5 X □) + (□ X 9) + 1 = (□ X 7) – 1

345. (5 X □) + (□ X 8) – 1 = (□ X 6) + 1

346. (5 X □) + (□ X 8) = □ X 6

347. (5 X □) + (□ X 9) – 1 = □ X 6

348. (5 X □) + (□ X 9) + 1 = □ X 7

349. (5 X □) + (□ X 9) + 1 = □ X 6

350. (5 X □) + (□ X 9) – 1 = □ X 7

351. (5 X □) + (□ X 6) – 1 = □ X 4

352. (5 X □) + (□ X 6) – 1 = (□ X 3) + 1

353. (6 X □) + (□ X 8) = □ X 9

354. (6 X □) + (□ X 8) = □ X 7

355. (6 X □) + (□ X 9) – 1 = (□ X 8) + 1

356. (6 X □) + (□ X 9) – 1 = □ X 8

357. (6 X □) + (□ X 8) – 1 = (□ X 7) + 1

358. (7 X □) + (□ X 9) = □ X 8

359. (7 X □) + (□ X 9) – 1 = (□ X 8) + 1

360. (7 X □) + (□ X 9) – 1 = □ X 8

You've just finishing another section.

Chapter Six   **Six-Solution Puzzles**

For the next ones, you are going to try to find six solutions for each puzzle.

361. $(2 \times \square) + (\square \times 8) + 1 = (\square \times 5) - 1$

362. $(2 \times \square) + (\square \times 8) - 1 = \square \times 3$

363. $(2 \times \square) + (\square \times 9) - 1 = (\square \times 3) + 1$

364. $(2 \times \square) + (\square \times 7) = \square \times 3$

365. $(2 \times \square) + (\square \times 6) = \square \times 3$

366. $(2 \times \square) + (\square \times 7) + 1 = (\square \times 3) - 1$

367. $(2 \times \square) + (\square \times 8) - 1 = (\square \times 3) + 1$

368. $(2 \times \square) + (\square \times 8) + 1 = \square \times 3$

369. $(2 \times \square) + (\square \times 9) = \square \times 6$

370. $(2 \times \square) + (\square \times 6) + 1 = (\square \times 3) - 1$

371. $(2 \times \square) + (\square \times 7) + 1 = (\square \times 4) - 1$

372. $(2 \times \square) + (\square \times 7) + 1 = \square \times 3$

373. $(2 \times \square) + (\square \times 9) + 1 = (\square \times 6) - 1$

374. $(2 \times \square) + (\square \times 9) + 1 = (\square \times 5) - 1$

375. $(2 \times \square) + (\square \times 9) + 1 = \square \times 5$

376. $(3 \times \square) + (\square \times 6) + 1 = \square \times 4$

377. $(3 \times \square) + (\square \times 7) + 1 = (\square \times 4) - 1$

378. $(3 \times \square) + (\square \times 8) + 1 = (\square \times 5) - 1$

379. $(3 \times \square) + (\square \times 8) - 1 = \square \times 5$

380. $(3 \times \square) + (\square \times 9) + 1 = (\square \times 8) - 1$

381. $(3 \times \square) + (\square \times 9) = \square \times 4$

382. $(3 \times \square) + (\square \times 6) - 1 = (\square \times 4) + 1$

383. $(3 \times \square) + (\square \times 6) + 1 = (\square \times 4) - 1$

384. $(3 \times \square) + (\square \times 7) + 1 = \square \times 4$

385. $(3 \times \square) + (\square \times 8) + 1 = (\square \times 6) - 1$

386. $(3 \times \square) + (\square \times 9) - 1 = (\square \times 7) + 1$

387. $(3 \times \square) + (\square \times 9) - 1 = \square \times 8$

388. $(3 \times \square) + (\square \times 7) = \square \times 4$

389. $(3 \times \square) + (\square \times 8) + 1 = \square \times 5$

390. $(3 \times \square) + (\square \times 7) - 1 = \square \times 4$

391. $(3 \times \square) + (\square \times 8) - 1 = (\square \times 4) + 1$

392. $(3 \times \square) + (\square \times 9) + 1 = (\square \times 7) - 1$

393. $(3 \times \square) + (\square \times 9) - 1 = (\square \times 5) + 1$

394. $(4 \times \square) + (\square \times 6) - 1 = \square \times 3$

395. $(4 \times \square) + (\square \times 8) + 1 = \square \times 7$

396. $(4 \times \square) + (\square \times 9) - 1 = \square \times 7$

397. $(4 \times \square) + (\square \times 5) - 1 = (\square \times 3) + 1$

398. $(4 \times \square) + (\square \times 6) + 1 = (\square \times 5) - 1$

399. $(4 \times \square) + (\square \times 7) = \square \times 6$

400. $(4 \times \square) + (\square \times 7) + 1 = \square \times 5$

401. $(4 \times \square) + (\square \times 7) - 1 = (\square \times 5) + 1$

402. $(4 \times \square) + (\square \times 7) + 1 = (\square \times 6) - 1$

403. $(4 \times \square) + (\square \times 7) = \square \times 5$

404. $(4 \times \square) + (\square \times 8) - 1 = (\square \times 9) + 1$

405. $(4 \times \square) + (\square \times 8) = \square \times 5$

406. $(4 \times \square) + (\square \times 8) - 1 = (\square \times 7) + 1$

407. $(4 \times \square) + (\square \times 8) + 1 = (\square \times 9) - 1$

408. $(4 \times \square) + (\square \times 9) + 1 = \square \times 8$

409. $(4 \times \square) + (\square \times 9) + 1 = \square \times 7$

410. $(4 \times \square) + (\square \times 9) + 1 = \square \times 6$

411. $(4 \times \square) + (\square \times 9) = \square \times 7$

412. $(4 \times \square) + (\square \times 9) + 1 = (\square \times 7) - 1$

413. $(5 \times \square) + (\square \times 7) + 1 = (\square \times 6) - 1$

414. $(5 \times \square) + (\square \times 8) + 1 = (\square \times 7) - 1$

415. $(5 \times \square) + (\square \times 8) + 1 = \square \times 6$

416. $(5 \times \square) + (\square \times 9) = \square \times 8$

417. $(5 \times \square) + (\square \times 9) = \square \times 7$

418. $(5 \times \square) + (\square \times 9) + 1 = \square \times 8$

419. $(5 \times \square) + (\square \times 9) + 1 = (\square \times 8) - 1$

420. $(5 \times \square) + (\square \times 7) + 1 = \square \times 6$

421. $(5 \times \square) + (\square \times 8) - 1 = \square \times 7$

422. $(5 \times \square) + (\square \times 8) = \square \times 7$

423. $(5 \times \square) + (\square \times 8) + 1 = \square \times 7$

424. $(5 \times \square) + (\square \times 9) - 1 = (\square \times 8) + 1$

425. $(5 \times \square) + (\square \times 9) - 1 = (\square \times 7) + 1$

426. $(5 \times \square) + (\square \times 9) - 1 = \square \times 8$

427. $(5 \times \square) + (\square \times 7) - 1 = (\square \times 6) + 1$

428. $(5 \times \square) + (\square \times 7) - 1 = \square \times 6$

429. $(5 \times \square) + (\square \times 8) - 1 = (\square \times 7) + 1$

430. $(5 \times \square) + (\square \times 8) - 1 = \square \times 6$

431. $(5 \times \square) + (\square \times 9) - 1 = (\square \times 6) + 1$

432. $(6 \times \square) + (\square \times 8) + 1 = (\square \times 9) - 1$

433. $(6 \times \square) + (\square \times 8) + 1 = \square \times 7$

434. $(6 \times \square) + (\square \times 9) = \square \times 8$

435. $(6 \times \square) + (\square \times 9) = \square \times 7$

436. $(6 \times \square) + (\square \times 9) + 1 = \square \times 8$

437. $(6 \times \square) + (\square \times 7) + 1 = (\square \times 8) - 1$

438. $(6 \times \square) + (\square \times 8) - 1 = \square \times 7$

439. $(6 \times \square) + (\square \times 7) = \square \times 9$

440. $(6 \times \square) + (\square \times 7) = \square \times 8$

441. $(6 \times \square) + (\square \times 7) - 1 = (\square \times 8) + 1$

442. $(7 \times \square) + (\square \times 8) + 1 = (\square \times 9) - 1$

443. $(7 \times \square) + (\square \times 8) + 1 = \square \times 9$

444. $(7 \times \square) + (\square \times 8) = \square \times 9$

445. $(7 \times \square) + (\square \times 8) - 1 = \square \times 9$

446. $(7 \times \square) + (\square \times 8) - 1 = (\square \times 9) + 1$

There's that minus sign again between the first two groups.

447. $(9 \text{ X } \square) - (\square \text{ X } 5) - 1 = \square \text{ X } 8$

448. $(9 \text{ X } \square) - (\square \text{ X } 6) = \square \text{ X } 7$

449. $(9 \text{ X } \square) - (\square \text{ X } 7) = \square \text{ X } 6$

450. $(9 \text{ X } \square) - (\square \text{ X } 7) - 1 = \square \text{ X } 8$

451. $(9 \text{ X } \square) - (\square \text{ X } 7) = \square \text{ X } 8$

452. $(9 \text{ X } \square) - (\square \text{ X } 7) + 1 = \square \text{ X } 8$

453. $(9 \text{ X } \square) - (\square \text{ X } 7) + 1 = (\square \text{ X } 8) - 1$

454. $(9 \text{ X } \square) - (\square \text{ X } 7) - 1 = (\square \text{ X } 8) + 1$

455. $(9 \text{ X } \square) - (\square \text{ X } 8) + 1 = (\square \text{ X } 7) - 1$

456. $(9 \text{ X } \square) - (\square \text{ X } 8) + 1 = \square \text{ X } 7$

457. $(9 \text{ X } \square) - (\square \text{ X } 8) = \square \text{ X } 7$

458. $(9 \text{ X } \square) - (\square \text{ X } 8) - 1 = \square \text{ X } 7$

459. $(9 \text{ X } \square) - (\square \text{ X } 8) - 1 = (\square \text{ X } 7) + 1$

460. $(8 \text{ X } \square) - (\square \text{ X } 5) = \square \text{ X } 9$

461. $(8 \text{ X } \square) - (\square \text{ X } 5) - 1 = \square \text{ X } 9$

462. $(8 \text{ X } \square) - (\square \text{ X } 7) = \square \text{ X } 6$

463. $(8 \text{ X } \square) - (\square \text{ X } 9) = \square \text{ X } 6$

464. $(8 \text{ X } \square) - (\square \text{ X } 9) = \square \text{ X } 5$

465. $(8 \text{ X } \square) - (\square \text{ X } 3) + 1 = \square \text{ X } 9$

You've just completed solving the puzzles with six solutions.

Chapter Seven   Seven-Solution Puzzles

For each puzzle in this section, try to find seven solutions. Even if you may not be able to find all seven, you may still pat yourself on the back for having found some, assuming you do. At least, you were a good sport in trying.

We're back to having a plus between the first two groups.

466. $(2 \, X \, \square) + (\square \, X \, 7) = \square \, X \, 7$

467. $(2 \, X \, \square) + (\square \, X \, 7) + 1 = (\square \, X \, 7) - 1$

468. $(2 \, X \, \square) + (\square \, X \, 3) + 1 = (\square \, X \, 2) - 1$

469. $(2 \, X \, \square) + (\square \, X \, 5) = \square \, X \, 5$

470. $(2 \, X \, \square) + (\square \, X \, 7) = \square \, X \, 4$

471. $(2 \, X \, \square) + (\square \, X \, 7) - 1 = \square \, X \, 3$

472. $(2 \, X \, \square) + (\square \, X \, 9) = \square \, X \, 8$

473. $(2 \, X \, \square) + (\square \, X \, 9) = \square \, X \, 7$

474. $(2 \, X \, \square) + (\square \, X \, 9) + 1 = (\square \, X \, 7) - 1$

475. $(2 \, X \, \square) + (\square \, X \, 9) + 1 = (\square \, X \, 8) - 1$

476. $(2 \, X \, \square) + (\square \, X \, 9) = \square \, X \, 5$

477. $(2 \, X \, \square) + (\square \, X \, 9) - 1 = (\square \, X \, 5) + 1$

478. $(2 \, X \, \square) + (\square \, X \, 7) - 1 = (\square \, X \, 4) + 1$

479. $(2 \, X \, \square) + (\square \, X \, 7) - 1 = (\square \, X \, 3) + 1$

480. $(2 \, X \, \square) + (\square \, X \, 9) - 1 = (\square \, X \, 8) + 1$

481. $(2 \, X \, \square) + (\square \, X \, 9) - 1 = \square \, X \, 4$

482. $(2 \, X \, \square) + (\square \, X \, 9) + 1 = (\square \, X \, 9) - 1$

483. $(2 \, X \, \square) + (\square \, X \, 8) = \square \, X \, 5$

484. $(2 \times \square) + (\square \times 9) - 1 = (\square \times 7) + 1$

485. $(2 \times \square) + (\square \times 9) - 1 = (\square \times 6) + 1$

486. $(2 \times \square) + (\square \times 9) - 1 = \square \times 5$

487. $(2 \times \square) + (\square \times 5) + 1 = (\square \times 3) - 1$

488. $(3 \times \square) + (\square \times 6) + 1 = (\square \times 5) - 1$

489. $(3 \times \square) + (\square \times 7) = \square \times 6$

490. $(3 \times \square) + (\square \times 7) + 1 = \square \times 5$

491. $(3 \times \square) + (\square \times 8) = \square \times 5$

492. $(3 \times \square) + (\square \times 8) + 1 = \square \times 4$

493. $(3 \times \square) + (\square \times 9) + 1 = \square \times 8$

494. $(3 \times \square) + (\square \times 9) = \square \times 5$

495. $(3 \times \square) + (\square \times 6) - 1 = \square \times 5$

496. $(3 \times \square) + (\square \times 7) + 1 = (\square \times 5) - 1$

497. $(3 \times \square) + (\square \times 8) = \square \times 9$

498. $(3 \times \square) + (\square \times 8) + 1 = \square \times 7$

499. $(3 \times \square) + (\square \times 8) + 1 = (\square \times 7) - 1$

500. $(3 \times \square) + (\square \times 8) - 1 = (\square \times 5) + 1$

501. $(3 \times \square) + (\square \times 9) - 1 = (\square \times 8) + 1$

502. $(3 \times \square) + (\square \times 7) - 1 = (\square \times 5) + 1$

503. $(3 \times \square) + (\square \times 4) - 1 = \square \times 2$

504. $(4 \times \square) + (\square \times 7) - 1 = \square \times 5$

505. $(4 \times \square) + (\square \times 9) - 1 = \square \times 6$

506. $(4 \times \square) + (\square \times 6) - 1 = (\square \times 5) + 1$

507. $(4 \times \square) + (\square \times 7) = \square \times 8$

508. $(4 \times \square) + (\square \times 7) - 1 = (\square \times 6) + 1$

509. $(4 \times \square) + (\square \times 7) = (\square \times 5) + 1$

510. $(4 \times \square) + (\square \times 8) - 1 = \square \times 9$

511. $(4 \times \square) + (\square \times 8) = \square \times 7$

512. $(4 \times \square) + (\square \times 8) + 1 = \square \times 5$

513. $(4 \times \square) + (\square \times 9) + 1 = (\square \times 8) - 1$

514. $(4 \times \square) + (\square \times 9) - 1 = (\square \times 7) + 1$

515. $(4 \times \square) + (\square \times 9) - 1 = (\square \times 8) + 1$

516. $(5 \times \square) + (\square \times 7) + 1 = \square \times 9$

517. $(5 \times \square) + (\square \times 8) + 1 = (\square \times 9) - 1$

518. $(5 \times \square) + (\square \times 8) + 1 = \square \times 9$

519. $(5 \times \square) + (\square \times 6) + 1 = (\square \times 8) - 1$

520. $(5 \times \square) + (\square \times 7) + 1 = (\square \times 8) - 1$

521. $(5 \times \square) + (\square \times 7) + 1 = \square \times 8$

522. $(5 \times \square) + (\square \times 8) = \square \times 9$

523. $(5 \times \square) + (\square \times 8) - 1 = \square \times 9$

524. $(5 \times \square) + (\square \times 8) - 1 = (\square \times 9) + 1$

525. $(5 \times \square) + (\square \times 6) + 1 = (\square \times 7) - 1$

526. $(5 \times \square) + (\square \times 6) + 1 = \square \times 7$

527. $(5 \times \square) + (\square \times 7) = \square \times 8$

528. $(5 \times \square) + (\square \times 7) = \square \times 6$

529. $(6 \times \square) + (\square \times 7) - 1 = (\square \times 9) + 1$

530. $(6 \times \square) + (\square \times 7) + 1 = \square \times 9$

531. $(6 \times \square) + (\square \times 7) + 1 = \square \times 8$

532. $(6 \times \square) + (\square \times 8) - 1 = \square \times 9$

533. $(6 \times \square) + (\square \times 8) - 1 = (\square \times 9) + 1$

534. $(6 \times \square) + (\square \times 7) - 1 = \square \times 8$

535. $(6 \times \square) + (\square \times 8) - 1 = (\square \times 4) + 1$

Notice the minus sign again.

536. $(9 \times \square) - (\square \times 3) - 1 = (\square \times 8) + 1$

537. $(9 \times \square) - (\square \times 3) = \square \times 8$

538. $(9 \times \square) - (\square \times 4) - 1 = \square \times 7$

539. $(9 \times \square) - (\square \times 5) - 1 = \square \times 7$

540. $(9 \times \square) - (\square \times 5) + 1 = \square \times 6$

541. $(9 \times \square) - (\square \times 5) + 1 = (\square \times 8) - 1$

542. $(9 \times \square) - (\square \times 5) = \square \times 8$

543. $(9 \times \square) - (\square \times 5) - 1 = (\square \times 8) + 1$

544. $(9 \times \square) - (\square \times 5) + 1 = \square \times 8$

545. $(9 \times \square) - (\square \times 6) + 1 = \square \times 8$

546. $(9 \times \square) - (\square \times 6) - 1 = \square \times 7$

547. $(9 \times \square) - (\square \times 7) - 1 = \square \times 6$

548. $(9 \times \square) - (\square \times 7) - 1 = \square \times 5$

549. $(9 \times \square) - (\square \times 7) + 1 = (\square \times 6) - 1$

550. $(9 \times \square) - (\square \times 8) = \square \times 3$

551. $(9 \times \square) - (\square \times 8) + 1 = \square \times 6$

552. $(9 \times \square) - (\square \times 8) - 1 = (\square \times 5) + 1$

553. $(9 \times \square) - (\square \times 8) = \square \times 5$

554. $(9 \times \square) - (\square \times 8) - 1 = \square \times 5$

555. $(9 \times \square) - (\square \times 2) = \square \times 9$

556. $(9 \times \square) - (\square \times 5) - 1 = \square \times 6$

557. $(9 \times \square) - (\square \times 4) + 1 = \square \times 7$

558. $(9 \times \square) - (\square \times 4) + 1 = \square \times 8$

559. $(9 \times \square) - (\square \times 8) + 1 = \square \times 5$

560. $(8 \times \square) - (\square \times 2) = \square \times 9$

561. $(8 \times \square) - (\square \times 4) = \square \times 7$

562. $(8 \times \square) - (\square \times 5) = \square \times 7$

563. $(8 \times \square) - (\square \times 5) - 1 = \square \times 7$

564. $(8 \times \square) - (\square \times 7) = \square \times 4$

565. $(8 \times \square) - (\square \times 7) + 1 = \square \times 6$

566. $(8 \times \square) - (\square \times 7) = \square \times 5$

That's all of the seven-solution problems.

Chapter Eight   Eight-Solution Puzzles

At this point in your puzzle solving, you probably are spinning through several possible solutions in seconds – all being done in your head. If that's true, then your mind is getting a real workout which should have similar results to blowing out all the carbon buildup in your car's engine. You should notice your mind working a lot better. Your thinking is quicker now than when you first started these puzzles. However, if you should be using a sheet of multiplication tables or your calculator, then you are not benefiting as much as you would be by trying to solve the puzzles in your head and without these crutches. Try to use your head alone to solve these puzzles. It won't be long and you'll notice a favorable improvement in your thinking speed.

In this section, try to find eight solutions for each puzzle.

567. $(2 \times \square) + 1 = \square \times 9$

568. $(2 \times \square) + (\square \times 9) + 1 = \square \times 6$

569. $(2 \times \square) + (\square \times 8) - 1 = (\square \times 2) + 1$

570. $(2 \times \square) + (\square \times 9) - 1 = \square \times 9$

571. $(2 \times \square) + (\square \times 7) - 1 = \square \times 7$

572. $(2 \times \square) + (\square \times 7) + 1 = \square \times 4$

573. $(2 \times \square) + (\square \times 9) + 1 = \square \times 9$

574. $(2 \times \square) + (\square \times 3) = \square \times 8$

575. $(2 \times \square) + (\square \times 7) - 1 = (\square \times 8) + 1$

576. $(2 \times \square) + (\square \times 7) + 1 = (\square \times 6) - 1$

577. $(2 \times \square) + (\square \times 8) - 1 = (\square \times 7) + 1$

578. $(2 \times \square) + (\square \times 8) - 1 = (\square \times 5) + 1$

579. $(2 \times \square) + (\square \times 8) = \square \times 7$

580. $(2 \times \square) + (\square \times 8) + 1 = \square \times 5$

581. $(2 \times \square) + (\square \times 8) + 1 = (\square \times 9) - 1$

582. $(2 \times \square) + (\square \times 8) - 1 = \square \times 5$

583. $(2 \times \square) + (\square \times 8) + 1 = (\square \times 7) - 1$

584. $(2 \times \square) + (\square \times 9) + 1 = \square \times 8$

585. $(2 \times \square) + (\square \times 9) - 1 = \square \times 7$

586. $(2 \times \square) + (\square \times 9) + 1 = \square \times 7$

587. $(2 \times \square) + (\square \times 7) = \square \times 6$

588. $(2 \times \square) + (\square \times 7) + 1 = (\square \times 5) - 1$

589. $(2 \times \square) + (\square \times 8) = \square \times 9$

590. $(2 \times \square) + (\square \times 9) - 1 = \square \times 6$

591. $(2 \times \square) + (\square \times 9) - 1 = \square \times 8$

592. $(3 \times \square) + (\square \times 7) - 1 = \square \times 5$

593. $(3 \times \square) + (\square \times 7) + 1 = \square \times 6$

594. $(3 \times \square) + (\square \times 7) - 1 = (\square \times 4) + 1$

595. $(3 \times \square) + (\square \times 8) - 1 = (\square \times 7) + 1$

596. $(3 \times \square) + (\square \times 8) - 1 = \square \times 7$

597. $(3 \times \square) + (\square \times 8) - 1 = \square \times 6$

598. $(3 \times \square) + (\square \times 8) + 1 = (\square \times 9) - 1$

599. $(3 \times \square) + (\square \times 8) = \square \times 7$

600. $(3 \times \square) + (\square \times 9) = \square \times 8$

601. $(3 \times \square) + (\square \times 9) = \square \times 7$

602. $(3 \times \square) + (\square \times 9) - 1 = \square \times 7$

603. $(3 \times \square) + (\square \times 5) + 1 = (\square \times 4) - 1$

604. $(3 \times \square) + (\square \times 6) + 1 = (\square \times 8) - 1$

605. $(3 \times \square) + (\square \times 6) + 1 = \square \times 7$

606. $(3 \times \square) + (\square \times 6) = \square \times 4$

607. $(3 \times \square) + (\square \times 7) = \square \times 9$

608. $(3 \times \square) + (\square \times 7) - 1 = (\square \times 6) + 1$

609. $(3 \times \square) + (\square \times 8) - 1 = \square \times 9$

610. $(3 \times \square) + (\square \times 8) - 1 = (\square \times 6) + 1$

611. $(3 \times \square) + (\square \times 5) + 1 = \square \times 4$

612. $(4 \times \square) + (\square \times 6) = \square \times 9$

613. $(4 \times \square) + (\square \times 6) + 1 = \square \times 5$

614. $(4 \times \square) + (\square \times 6) - 1 = \square \times 5$

615. $(4 \times \square) + (\square \times 6) + 1 = (\square \times 7) - 1$

616. $(4 \times \square) + (\square \times 7) + 1 = \square \times 6$

617. $(4 \times \square) + (\square \times 7) - 1 = \square \times 9$

618. $(4 \times \square) + (\square \times 7) - 1 = \square \times 8$

619. $(4 \times \square) + (\square \times 7) - 1 = \square \times 6$

620. $(4 \times \square) + (\square \times 7) + 1 = (\square \times 9) - 1$

621. $(4 \times \square) + (\square \times 7) = \square \times 9$

622. $(4 \times \square) + (\square \times 7) + 1 = (\square \times 8) - 1$

623. $(4 \times \square) + (\square \times 8) = \square \times 9$

624. $(4 \times \square) + (\square \times 8) - 1 = \square \times 7$

625. $(4 \times \square) + (\square \times 9) - 1 = \square \times 8$

626. $(5 \times \square) + (\square \times 6) = \square \times 9$

627. $(5 \times \square) + (\square \times 6) = \square \times 7$

628. $(5 \times \square) + (\square \times 7) - 1 = \square \times 9$

629. $(5 \times \square) + (\square \times 7) - 1 = (\square \times 8) + 1$

630. $(5 \times \square) + (\square \times 7) + 1 = (\square \times 9) - 1$

631. $(5 \times \square) + (\square \times 6) + 1 = (\square \times 9) - 1$

632. $(5 \times \square) + (\square \times 6) = \square \times 8$

633. $(5 \times \square) + (\square \times 6) - 1 = (\square \times 8) + 1$

634. $(5 \times \square) + (\square \times 7) = \square \times 9$

635. $(5 \times \square) + (\square \times 7) - 1 = (\square \times 9) + 1$

636. $(5 \times \square) + (\square \times 6) + 1 = \square \times 9$

637. $(6 \times \square) + (\square \times 7) - 1 = \square \times 9$

638. $(6 \times \square) + (\square \times 7) + 1 = (\square \times 9) - 1$

639. $(6 \times \square) + (\square \times 8) + 1 = \square \times 9$

We're back to showing a minus sign again between the first two groups.

640. $(9 \times \square) - (\square \times 3) = \square \times 7$

641. $(9 \times \square) - (\square \times 3) + 1 = \square \times 8$

642. $(9 \times \square) - (\square \times 4) = \square \times 7$

643. $(9 \times \square) - (\square \times 5) = \square \times 6$

644. $(9 \times \square) - (\square \times 5) + 1 = \square \times 7$

645. $(9 \times \square) - (\square \times 5) - 1 = (\square \times 7) + 1$

646. $(9 \times \square) - (\square \times 5) = \square \times 7$

647. $(9 \times \square) - (\square \times 5) + 1 = (\square \times 7) - 1$

49

648. (9 X □) – (□ X 6) + 1 = □ X 7

649. (9 X □) – (□ X 6) – 1 = □ X 8

650. (9 X □) – (□ X 6) = □ X 5

651. (9 X □) – (□ X 6) – 1 = (□ X 7) + 1

652. (9 X □) – (□ X 7) = □ X 5

653. (9 X □) – (□ X 7) – 1 = (□ X 6) + 1

654. (9 X □) – (□ X 7) + 1 = □ X 5

655. (9 X □) – (□ X 7) + 1 = □ X 6

656. (9 X □) – (□ X 8) – 1 = □ X 6

657. (9 X □) – (□ X 2) = □ X 8

658. (9 X □) – (□ X 3) – 1 = □ X 8

659. (9 X □) – (□ X 4) = □ X 6

660. (9 X □) – (□ X 4) = □ X 8

661. (9 X □) – (□ X 5) = □ X 3

662. (9 X □) – (□ X 6) = □ X 4

663. (9 X □) – (□ X 7) = □ X 3

664. (9 X □) – (□ X 8) = □ X 4

665. (9 X □) – (□ X 6) – 1 = □ X 5

666. (9 X □) – (□ X 7) = □ X 4

667. (8 X □) – (□ X 2) + 1 = □ X 9

668. (8 X □) – (□ X 3) = □ X 9

669. (8 X □) – (□ X 4) + 1 = □ X 9

670. $(8 \times \square) - (\square \times 7) + 1 = \square \times 5$

671. $(8 \times \square) - (\square \times 9) = \square \times 3$

672. $(8 \times \square) - (\square \times 3) - 1 = \square \times 7$

You have now finished discovering eight different solutions for each of the above puzzles and are prepared to tackle the next chapter where you'll be trying to find nine solutions for each puzzle.

Chapter Nine   Nine-Solution Puzzles

673. $(2 \times \square) + (\square \times 3) = \square \times 9$

674. $(2 \times \square) + (\square \times 3) - 1 = (\square \times 8) + 1$

675. $(2 \times \square) + (\square \times 3) + 1 = (\square \times 8) - 1$

676. $(2 \times \square) + (\square \times 3) - 1 = (\square \times 9) + 1$

677. $(2 \times \square) + (\square \times 3) + 1 = \square \times 9$

678. $(2 \times \square) + (\square \times 4) = \square \times 9$

679. $(2 \times \square) + (\square \times 4) - 1 = \square \times 9$

680. $(2 \times \square) + (\square \times 4) - 1 = (\square \times 9) + 1$

681. $(2 \times \square) + (\square \times 4) + 1 = (\square \times 9) - 1$

682. $(2 \times \square) + (\square \times 4) + 1 = \square \times 9$

683. $(2 \times \square) + (\square \times 4) - 1 = (\square \times 2) + 1$

684. $(2 \times \square) + (\square \times 5) = \square \times 9$

685. $(2 \times \square) + (\square \times 5) - 1 = (\square \times 8) + 1$

686. $(2 \times \square) + (\square \times 5) + 1 = \square \times 9$

687. $(2 \times \square) + (\square \times 5) = \square \times 8$

688. $(2 \times \square) + (\square \times 5) - 1 = (\square \times 9) + 1$

689. $(2 \times \square) + (\square \times 5) + 1 = (\square \times 8) - 1$

690. $(2 \times \square) + (\square \times 5) - 1 = \square \times 9$

691. $(2 \times \square) + (\square \times 5) + 1 = \square \times 3$

692. $(2 \times \square) + (\square \times 6) = \square \times 9$

693. $(2 \times \square) + (\square \times 6) - 1 = (\square \times 3) + 1$

694. $(2 \times \square) + (\square \times 6) - 1 = (\square \times 9) + 1$

695. $(2 \times \square) + (\square \times 6) + 1 = \square \times 3$

696. $(2 \times \square) + (\square \times 7) - 1 = (\square \times 9) + 1$

697. $(2 \times \square) + (\square \times 7) + 1 = \square \times 9$

698. $(2 \times \square) + (\square \times 7) - 1 = \square \times 5$

699. $(2 \times \square) + (\square \times 7) + 1 = (\square \times 8) - 1$

700. $(2 \times \square) + (\square \times 7) - 1 = (\square \times 6) + 1$

701. $(2 \times \square) + (\square \times 7) - 1 = (\square \times 5) + 1$

702. $(2 \times \square) + (\square \times 7) = \square \times 8$

703. $(2 \times \square) + (\square \times 7) = \square \times 5$

704. $(2 \times \square) + (\square \times 7) + 1 = (\square \times 9) - 1$

705. $(2 \times \square) + (\square \times 7) + 1 = \square \times 5$

706. $(2 \times \square) + (\square \times 8) - 1 = \square \times 9$

707. $(2 \times \square) + (\square \times 8) - 1 = (\square \times 9) + 1$

708. $(2 \times \square) + (\square \times 8) - 1 = \square \times 7$

709. $(2 \times \square) + (\square \times 8) + 1 = \square \times 7$

710. $(2 \times \square) + (\square \times 8) + 1 = \square \times 9$

711. $(2 \times \square) + (\square \times 3) + 1 = (\square \times 9) - 1$

712. $(2 \times \square) + (\square \times 3) - 1 = \square \times 9$

713. $(2 \times \square) + (\square \times 5) + 1 = (\square \times 4) - 1$

714. $(2 \times \square) + (\square \times 6) + 1 = (\square \times 9) - 1$

715. $(2 \times \square) + (\square \times 6) - 1 = \square \times 9$

716. $(2 \times \square) + (\square \times 6) - 1 = \square \times 3$

717. $(2 \times \square) + (\square \times 6) + 1 = (\square \times 5) - 1$

718. $(3 \times \square) + (\square \times 4) = \square \times 4$

719. $(3 \times \square) + (\square \times 4) = \square \times 9$

720. $(3 \times \square) + (\square \times 4) - 1 = \square \times 9$

721. $(3 \times \square) + (\square \times 4) - 1 = (\square \times 9) + 1$

722. $(3 \times \square) + (\square \times 4) + 1 = \square \times 9$

723. $(3 \times \square) + (\square \times 5) = \square \times 9$

724. $(3 \times \square) + (\square \times 5) + 1 = (\square \times 9) - 1$

725. $(3 \times \square) + (\square \times 5) - 1 = (\square \times 9) + 1$

726. $(3 \times \square) + (\square \times 5) + 1 = \square \times 8$

727. $(3 \times \square) + (\square \times 5) - 1 = \square \times 9$

728. $(3 \times \square) + (\square \times 6) + 1 = \square \times 5$

729. $(3 \times \square) + (\square \times 6) - 1 = \square \times 4$

730. $(3 \times \square) + (\square \times 7) - 1 = (\square \times 9) + 1$

731. $(3 \times \square) + (\square \times 7) + 1 = \square \times 9$

732. $(3 \times \square) + (\square \times 7) + 1 = \square \times 8$

733. $(3 \times \square) + (\square \times 7) = \square \times 8$

734. $(3 \times \square) + (\square \times 7) - 1 = \square \times 8$

735. $(3 \times \square) + (\square \times 7) + 1 = (\square \times 9) - 1$

736. $(3 \times \square) + (\square \times 8) + 1 = \square \times 9$

737. $(3 \times \square) + (\square \times 8) + 1 = \square \times 6$

738. $(3 \times \square) + (\square \times 4) = \square \times 8$

739. $(3 \times \square) + (\square \times 6) = \square \times 8$

740. $(3 \times \square) + (\square \times 6) - 1 = (\square \times 8) + 1$

741. $(3 \times \square) + (\square \times 6) - 1 = (\square \times 5) + 1$

742. $(3 \times \square) + (\square \times 7) + 1 = (\square \times 8) - 1$

743. $(3 \times \square) + (\square \times 4) - 1 = \square \times 8$

744. $(3 \times \square) + (\square \times 4) + 1 = (\square \times 8) - 1$

745. $(3 \times \square) + (\square \times 4) - 1 = (\square \times 8) + 1$

746. $(4 \times \square) + (\square \times 5) + 1 = \square \times 9$

747. $(4 \times \square) + (\square \times 5) = \square \times 8$

748. $(4 \times \square) + (\square \times 5) = \square \times 6$

749. $(4 \times \square) + (\square \times 5) + 1 = (\square \times 9) - 1$

750. $(4 \times \square) + (\square \times 5) - 1 = (\square \times 9) + 1$

751. $(4 \times \square) + (\square \times 5) + 1 = (\square \times 8) - 1$

752. $(4 \times \square) + (\square \times 5) - 1 = \square \times 9$

753. $(4 \times \square) + (\square \times 5) + 1 = (\square \times 6) - 1$

754. $(4 \times \square) + (\square \times 5) + 1 = \square \times 8$

755. $(4 \times \square) + (\square \times 5) - 1 = (\square \times 8) + 1$

756. $(4 \times \square) + (\square \times 6) + 1 = \square \times 9$

757. $(4 \times \square) + (\square \times 6) - 1 = (\square \times 9) + 1$

758. $(4 \times \square) + (\square \times 6) = \square \times 7$

759. $(4 \times \square) + (\square \times 6) - 1 = \square \times 7$

760. $(4 \times \square) + (\square \times 6) - 1 = (\square \times 7) + 1$

761. $(4 \times \square) + (\square \times 7) + 1 = \square \times 9$

762. $(4 \times \square) + (\square \times 7) - 1 = (\square \times 8) + 1$

763. $(4 \times \square) + (\square \times 8) + 1 = \square \times 9$

764. $(4 \times \square) + (\square \times 6) + 1 = (\square \times 9) - 1$

765. $(5 \times \square) + (\square \times 6) + 1 = \square \times 8$

766. $(5 \times \square) + (\square \times 6) - 1 = \square \times 8$

767. $(5 \times \square) + (\square \times 6) - 1 = \square \times 7$

768. $(5 \times \square) + (\square \times 6) - 1 = (\square \times 7) + 1$

769. $(5 \times \square) + (\square \times 7) - 1 = \square \times 8$

770. $(5 \times \square) + (\square \times 6) - 1 = \square \times 9$

771. $(5 \times \square) + (\square \times 6) - 1 = (\square \times 9) + 1$

Yes, there is that minus sign again between the first two groups. Nevertheless, you are still finding nine solutions for each puzzle.

772. $(9 \times \square) - (\square \times 2) - 1 = \square \times 8$

773. $(9 \times \square) - (\square \times 3) + 1 = (\square \times 7) - 1$

774. $(9 \times \square) - (\square \times 3) - 1 = \square \times 8$

775. $(9 \times \square) - (\square \times 4) - 1 = \square \times 6$

776. $(9 \times \square) - (\square \times 4) - 1 = \square \times 8$

777. $(9 \times \square) - (\square \times 8) - 1 = \square \times 4$

778. $(9 \times \square) - (\square \times 8) - 1 = \square \times 3$

779. $(9 \times \square) - (\square \times 2) + 1 = \square \times 7$

780. $(9 \times \square) - (\square \times 2) + 1 = (\square \times 7) - 1$

781. (9 X □) – (□ X 3) = □ X 5

782. (9 X □) – (□ X 3) – 1 = (□ X 7) + 1

783. (9 X □) – (□ X 4) + 1 = (□ X 6) – 1

784. (9 X □) – (□ X 4) + 1 = □ X 5

785. (9 X □) – (□ X 6) = □ X 2

786. (9 X □) – (□ X 7) – 1 = □ X 4

787. (9 X □) – (□ X 2) = □ X 5

788. (9 X □) – (□ X 2) + 1 = □ X 6

789. (9 X □) – (□ X 3) = □ X 4

790. (9 X □) – (□ X 4) = □ X 3

791. (9 X □) – (□ X 6) + 1 = □ X 5

792. (9 X □) – (□ X 2) = □ X 6

793. (9 X □) – (□ X 2) + 1 = (□ X 8) – 1

794. (9 X □) – (□ X 2) + 1 = □ X 5

795. (9 X □) – (□ X 3) + 1 = □ X 4

796. (9 X □) – (□ X 4) + 1 = □ X 3

797. (9 X □) – (□ X 4) – 1 = □ X 5

798. (9 X □) – (□ X 5) + 1 = □ X 4

799. (9 X □) – (□ X 2) = □ X 3

800. (9 X □) – (□ X 2) = □ X 7

801. (9 X □) – (□ X 3) + 1 = □ X 7

802. (9 X □) – (□ X 4) = □ X 3

803. (9 X □) – (□ X 2) = □ X 5

804. (9 X □) – (□ X 2) – 1 = □ X 7

805. (9 X □) – (□ X 2) – 1 = □ X 5

806. (9 X □) – (□ X 2) = □ X 4

807. (9 X □) – (□ X 2) + 1 = □ X 8

808. (8 X □) – (□ X 2) + 1 = (□ X 7) – 1

809. (8 X □) – (□ X 2) = □ X 7

810. (8 X □) – (□ X 3) + 1 = □ X 7

811. (8 X □) – (□ X 3) = □ X 6

812. (8 X □) – (□ X 3) – 1 = □ X 7

813. (8 X □) – (□ X 3) = □ X 7

814. (8 X □) – (□ X 4) = □ X 5

815. (8 X □) – (□ X 5) = □ X 4

816. (8 X □) – (□ X 5) – 1 = □ X 6

817. (8 X □) – (□ X 5) + 1 = □ X 6

818. (8 X □) – (□ X 3) = □ X 4

819. (8 X □) – (□ X 3) = □ X 7

820. (8 X □) – (□ X 3) + 1 = □ X 4

821. (8 X □) – (□ X 3) + 1 = □ X 7

822. (8 X □) – (□ X 3) – 1 = □ X 7

That concludes the section of nine-solution puzzles. In the next chapter, try to find as many as you can, up to ten solutions for each puzzle.

Chapter Ten   **Ten-Solution Puzzles**

823. $(2 \times \square) + (\square \times 4) + 1 = (\square \times 3) - 1$

824. $(2 \times \square) + (\square \times 5) - 1 = (\square \times 3) + 1$

825. $(2 \times \square) + (\square \times 5) - 1 = \square \times 3$

826. $(2 \times \square) + (\square \times 6) + 1 = (\square \times 7) - 1$

827. $(2 \times \square) + (\square \times 6) - 1 = (\square \times 5) + 1$

828. $(2 \times \square) + (\square \times 6) = \square \times 5$

829. $(2 \times \square) + (\square \times 7) - 1 = \square \times 4$

830. $(2 \times \square) + (\square \times 7) + 1 = \square \times 6$

831. $(2 \times \square) + (\square \times 7) - 1 = \square \times 6$

832. $(2 \times \square) + (\square \times 3) = \square \times 3$

833. $(2 \times \square) + (\square \times 3) + 1 = (\square \times 6) - 1$ [Actually, this one has twelve solutions.]

834. $(2 \times \square) + (\square \times 5) = \square \times 3$

835. $(3 \times \square) + (\square \times 5) + 1 = (\square \times 8) - 1$

836. $(3 \times \square) + (\square \times 5) - 1 = \square \times 8$

837. $(3 \times \square) + (\square \times 5) - 1 = (\square \times 8) + 1$

838. $(3 \times \square) + (\square \times 5) - 1 = \square \times 4$

839. $(3 \times \square) + (\square \times 5) = \square \times 4$

840. $(3 \times \square) + (\square \times 6) + 1 = (\square \times 7) - 1$

841. $(3 \times \square) + (\square \times 7) - 1 = \square \times 6$

842. $(3 \times \square) + (\square \times 7) + 1 = (\square \times 6) - 1$

843. $(3 \times \square) + (\square \times 5) - 1 = (\square \times 4) + 1$

844. $(4 \times \square) + (\square \times 5) + 1 = \square \times 7$

845. $(4 \times \square) + (\square \times 5) - 1 = \square \times 7$

846. $(4 \times \square) + (\square \times 5) - 1 = (\square \times 6) + 1$

847. $(4 \times \square) + (\square \times 5) + 1 = (\square \times 7) - 1$

848. $(4 \times \square) + (\square \times 6) + 1 = \square \times 7$

849. $(4 \times \square) + (\square \times 7) + 1 = \square \times 8$

Each of the next two puzzles has a minus sign between the first two groups.

850. $(8 \times \square) - (\square \times 3) + 1 = \square \times 5$

851. $(8 \times \square) - (\square \times 3) - 1 = \square \times 5$

You've now completed the puzzles with ten solutions.

Chapter Eleven   Eleven-Solution Puzzles

For this chapter, you'll want to try to find eleven solutions for each puzzle.

852. $(2 \times \square) + (\square \times 3) + 1 = \square \times 8$

853. $(2 \times \square) + (\square \times 3) - 1 = \square \times 8$

854. $(2 \times \square) + (\square \times 3) - 1 = (\square \times 7) + 1$

855. $(2 \times \square) + (\square \times 3) + 1 = \square \times 7$

856. $(2 \times \square) + (\square \times 4) = \square \times 7$

857. $(2 \times \square) + (\square \times 4) = \square \times 3$

858. $(2 \times \square) + (\square \times 4) - 1 = (\square \times 7) + 1$

859. $(2 \times \square) + (\square \times 5) + 1 = \square \times 7$

860. $(2 \times \square) + (\square \times 5) - 1 = \square \times 7$

861. $(2 \times \square) + (\square \times 5) - 1 = (\square \times 4) + 1$

862. $(2 \times \square) + (\square \times 5) - 1 = \square \times 8$

863. $(2 \times \square) + (\square \times 5) = \square \times 4$

864. $(2 \times \square) + (\square \times 5) + 1 = (\square \times 6) - 1$

865. $(2 \times \square) + (\square \times 6) + 1 = \square \times 7$

866. $(2 \times \square) + (\square \times 6) = \square \times 7$

867. $(2 \times \square) + (\square \times 6) - 1 = \square \times 5$

868. $(2 \times \square) + (\square \times 6) - 1 = (\square \times 7) + 1$

869. $(2 \times \square) + (\square \times 6) + 1 = \square \times 5$

870. $(2 \times \square) + (\square \times 8) = \square \times 4$

871. $(3 \times \square) + (\square \times 4) + 1 = \square \times 7$

872. $(3 \times \square) + (\square \times 4) - 1 = (\square \times 7) + 1$

873. $(3 \times \square) + (\square \times 4) + 1 = (\square \times 7) - 1$

874. $(3 \times \square) + (\square \times 4) - 1 = \square \times 7$

875. $(3 \times \square) + (\square \times 5) + 1 = \square \times 7$

876. $(3 \times \square) + (\square \times 5) + 1 = (\square \times 6) - 1$

877. $(3 \times \square) + (\square \times 5) + 1 = (\square \times 7) - 1$

878. $(3 \times \square) + (\square \times 5) = \square \times 7$

879. $(3 \times \square) + (\square \times 5) = \square \times 6$

880. $(3 \times \square) + (\square \times 5) - 1 = (\square \times 7) + 1$

881. $(3 \times \square) + (\square \times 5) - 1 = (\square \times 5) + 1$

882. $(3 \times \square) + (\square \times 6) = \square \times 7$

883. $(3 \times \square) + (\square \times 6) + 1 = \square \times 8$

884. $(3 \times \square) + (\square \times 6) - 1 = \square \times 7$

885. $(3 \times \square) + (\square \times 6) = \square \times 5$

886. $(3 \times \square) + (\square \times 4) = \square \times 6$

887. $(4 \times \square) + (\square \times 5) + 1 = \square \times 6$

888. $(4 \times \square) + (\square \times 5) - 1 = \square \times 6$

889. $(4 \times \square) + (\square \times 5) = \square \times 7$

You've just finished solving the eleven-solution puzzles. Now, see if you can find twelve solutions for each puzzle in the next chapter.

Chapter Twelve   Twelve-Solution Puzzles

890. $(2 \times \square) + (\square \times 3) - 1 = (\square \times 6) + 1$

891. $(2 \times \square) + (\square \times 3) - 1 = (\square \times 2) + 1$

892. $(2 \times \square) + (\square \times 3) + 1 = \square \times 2$

893. $(2 \times \square) + (\square \times 3) = \square \times 7$

894. $(2 \times \square) + (\square \times 3) = \square \times 6$

895. $(2 \times \square) + (\square \times 3) - 1 = \square \times 7$

896. $(2 \times \square) + (\square \times 4) + 1 = (\square \times 7) - 1$

897. $(2 \times \square) + (\square \times 4) - 1 = \square \times 7$

898. $(2 \times \square) + (\square \times 4) + 1 = (\square \times 4) - 1$

899. $(2 \times \square) + (\square \times 4) + 1 = (\square \times 2) - 1$

900. $(2 \times \square) + (\square \times 5) - 1 = (\square \times 7) + 1$

901. $(2 \times \square) + (\square \times 5) = \square \times 6$

902. $(2 \times \square) + (\square \times 5) - 1 = (\square \times 6) + 1$

903. $(2 \times \square) + (\square \times 5) + 1 = (\square \times 7) - 1$

904. $(2 \times \square) + (\square \times 5) + 1 = \square \times 4$

905. $(2 \times \square) + (\square \times 5) + 1 = (\square \times 5) - 1$

906. $(3 \times \square) + (\square \times 4) + 1 = (\square \times 5) - 1$

907. $(3 \times \square) + (\square \times 5) + 1 = \square \times 6$

908. $(3 \times \square) + (\square \times 5) - 1 = \square \times 6$

909. $(3 \times \square) + (\square \times 4) - 1 = (\square \times 6) + 1$

910. $(3 \times \square) + (\square \times 4) + 1 = (\square \times 6) - 1$

Chapter Thirteen   **Puzzles with Multiple Solutions**

A. Thirteen-Solution Puzzles

See what you can do now with the thirteen-solution puzzles in this next section.

911. $(2 \text{ X } \square) + (\square \text{ X } 4) - 1 = (\square \text{ X } 3) + 1$

912. $(2 \text{ X } \square) + (\square \text{ X } 4) + 1 = \square \text{ X } 3$

913. $(2 \text{ X } \square) + (\square \text{ X } 5) - 1 = \square \text{ X } 4$

914. $(3 \text{ X } \square) + (\square \text{ X } 4) + 1 = \square \text{ X } 5$

915. $(3 \text{ X } \square) + (\square \text{ X } 4) - 1 = \square \text{ X } 5$

916. $(4 \text{ X } \square) + (\square \text{ X } 8) + 1 = (\square \text{ X } 6) - 1$

That was quick!

B. Fourteen-Solution Puzzles

In this section, try finding fourteen solutions for each puzzle.

917. $(2 \text{ X } \square) + (\square \text{ X } 4) + 1 = (\square \text{ X } 5) - 1$

918. $(2 \text{ X } \square) + (\square \text{ X } 4) - 1 = \square \text{ X } 3$

919. $(2 \text{ X } \square) + (\square \text{ X } 5) + 1 = \square \text{ X } 6$

920. $(2 \text{ X } \square) + (\square \text{ X } 5) - 1 = \square \text{ X } 5$

921. $(2 \text{ X } \square) + (\square \text{ X } 8) + 1 = (\square \text{ X } 4) - 1$

922. $(3 \text{ X } \square) + (\square \text{ X } 4) + 1 = \square \text{ X } 6$

923. $(3 \text{ X } \square) + (\square \text{ X } 4) = \square \text{ X } 5$

924. $(4 \text{ X } \square) + (\square \text{ X } 8) - 1 = (\square \text{ X } 6) + 1$

That's the end of the fourteen-solution puzzles.

C. Fifteen-Solution Puzzles

925. $(2 \times \square) + (\square \times 3) + 1 = (\square \times 5) - 1$

926. $(2 \times \square) + (\square \times 4) - 1 = (\square \times 5) + 1$

927. $(2 \times \square) + (\square \times 4) + 1 = \square \times 5$

928. $(2 \times \square) + (\square \times 4) = \square \times 5$

929. $(2 \times \square) + (\square \times 3) - 1 = \square \times 6$

930. $(2 \times \square) + (\square \times 8) + 1 = (\square \times 6) - 1$

931. $(2 \times \square) + (\square \times 8) + 1 = (\square \times 8) - 1$

932. $(2 \times \square) + (\square \times 3) + 1 = (\square \times 3) - 1$

933. $(2 \times \square) + (\square \times 3) - 1 = \square \times 2$

934. $(3 \times \square) + (\square \times 4) + 1 = \square \times 4$

For each puzzle in the next section, try to find sixteen different solutions.

D. Sixteen-Solution Puzzles

935. $(2 \times \square) + (\square \times 3) - 1 = (\square \times 5) + 1$

936. $(2 \times \square) + (\square \times 3) + 1 = \square \times 5$

937. $(2 \times \square) + (\square \times 3) + 1 = (\square \times 4) - 1$

938. $(2 \times \square) + (\square \times 8) = \square \times 6$

939. $(2 \times \square) + (\square \times 8) - 1 = (\square \times 4) + 1$

Now, you are going to be trying to find seventeen solutions for each of the following puzzles.

E. Seventeen-Solution Puzzles

940. $(2 \text{ X } \square) + (\square \text{ X } 3) - 1 = \square \text{ X } 5$

941. $(2 \text{ X } \square) + (\square \text{ X } 3) = \square \text{ X } 4$

942. $(2 \text{ X } \square) + (\square \text{ X } 3) - 1 = (\square \text{ X } 4) + 1$

943. $(2 \text{ X } \square) + (\square \text{ X } 6) + 1 = (\square \text{ X } 4) - 1$

944. $(2 \text{ X } \square) + (\square \text{ X } 8) - 1 = (\square \text{ X } 6) + 1$

945. $(4 \text{ X } \square) + (\square \text{ X } 6) = \square \text{ X } 8$

For the last puzzle to find seventeen solutions, you'll notice a minus sign between the first two groups.

946. $(8 \text{ X } \square) - (\square \text{ X } 4) = \square \text{ X } 6$

F. Eighteen-Solution Puzzles

You now have only two puzzles in which to find eighteen solutions for each one.

947. $(2 \text{ X } \square) + (\square \text{ X } 3) - 1 = \square \text{ X } 3$

948. $(2 \text{ X } \square) + (\square \text{ X } 4) = \square \text{ X } 8$

G. Nineteen-Solution Puzzles

For each of the next four puzzles, try to find nineteen different solutions.

949. $(2 \text{ X } \square) + (\square \text{ X } 3) + 1 = \square \text{ X } 4$

950. $(2 \text{ X } \square) + (\square \text{ X } 6) = \square \text{ X } 4$

951. $(3 \text{ X } \square) + (\square \text{ X } 9) = \square \text{ X } 6$

952. $(4 \text{ X } \square) + (\square \text{ X } 6) + 1 = (\square \text{ X } 8) - 1$

H. Twenty-Solution Puzzles

In this section of three puzzles, you'll be finding twenty solutions for each one.

953. $(2 \times \square) + (\square \times 6) - 1 = (\square \times 8) + 1$

954. $(2 \times \square) + (\square \times 6) + 1 = (\square \times 8) - 1$

955. $(2 \times \square) + (\square \times 6) - 1 = (\square \times 4) + 1$

I. Twenty-One-Solution Puzzle

See how many of the twenty-one different solutions you can find in this next puzzle.

956. $(2 \times \square) + (\square \times 6) + 1 = (\square \times 6) - 1$

J. Twenty-Two-Solution Puzzle

Now try to find twenty-two unique solutions for this one.

957. $(2 \times \square) + (\square \times 4) - 1 = (\square \times 8) + 1$

K. Twenty-Seven-Solution Puzzles

With the next two puzzles you'll be trying to find twenty-seven different solutions for each.

958. $(2 \times \square) + (\square \times 4) - 1 = (\square \times 6) + 1$

959. $(2 \times \square) + (\square \times 4) + 1 = (\square \times 6) - 1$

Chapter Fourteen   Randomly Mixed Puzzles with Varying Numbers of Solutions

In this chapter, the puzzles appear in no particular order. To the right of each puzzle will be a number in parenthesis representing the number of possible solutions for that puzzle. Also, notice the signs as there may be a plus or a minus sign between the first two groups.

960. $6 \times \square = \square \times 9$ (3 Solutions)

961. $(2 \times \square) + (\square \times 9) + 1 = \square \times 3$ (3)

962. $(3 \times \square) + (\square \times 9) + 1 = \square \times 4$ (3)

963. $4 \times \square = \square \times 6$ (3)

964. $(3 \times \square) + (\square \times 6) = \square \times 2$ (3)

965. $(3 \times \square) + (\square \times 7) - 1 = (\square \times 2) + 1$ (3)

966. $(4 \times \square) + (\square \times 9) + 1 = (\square \times 3) - 1$ (3)

967. $(4 \times \square) + (\square \times 7) + 1 = (\square \times 3) - 1$ (3)

968. $(7 \times \square) + (\square \times 8) = \square \times 6$ (3)

969. $(2 \times \square) - 1 = (\square \times 4) + 1$ (4)

970. $(3 \times \square) + (\square \times 9) + 1 = (\square \times 4) - 1$ (4)

971. $(2 \times \square) + (\square \times 6) = (\square \times 9) + 1$ (9)

972. $(7 \times \square) - (\square \times 9) = \square \times 3$ (8)

973. $(7 \times \square) - (\square \times 9) + 1 = \square \times 5$ (5)

974. $(7 \times \square) - (\square \times 9) - 1 = \square \times 8$ (3)

975. $(7 \times \square) - (\square \times 8) - 1 = \square \times 6$ (6)

976. $(7 \times \square) - (\square \times 6) + 1 = \square \times 5$ (9)

977. $(7 \times \square) - (\square \times 5) + 1 = (\square \times 4) - 1$ (11)

978. $(7 \times \square) - (\square \times 4) + 1 = \square \times 8$ (8)

979. $(7 \times \square) - (\square \times 3) - 1 = \square \times 5$ (11)

980. $(7 \times \square) - (\square \times 2) = \square \times 9$ (7)

981. $(6 \times \square) - (\square \times 9) + 1 = \square \times 4$ (7)

982. $(6 \times \square) - (\square \times 8) - 1 = \square \times 9$ (4)

983. $(6 \times \square) - (\square \times 7) + 1 = \square \times 8$ (4)

984. $(6 \times \square) - (\square \times 5) - 1 = \square \times 8$ (6)

985. $(6 \times \square) - (\square \times 4) - 1 = (\square \times 8) + 1$ (13)

986. $(6 \times \square) - (\square \times 3) + 1 = \square \times 7$ (10)

987. $(6 \times \square) - (\square \times 2) - 1 = \square \times 7$ (10)

988. $(5 \times \square) - (\square \times 9) - 1 = \square \times 7$ (3)

989. $(5 \times \square) - (\square \times 8) - 1 = \square \times 9$ (2)

990. $(5 \times \square) - (\square \times 7) - 1 = (\square \times 9) + 1$ (2)

991. $(5 \times \square) - (\square \times 6) - 1 = (\square \times 7) + 1$ (4)

992. $(5 \times \square) - (\square \times 4) + 1 = \square \times 7$ (7)

993. $(5 \times \square) - (\square \times 3) + 1 = \square \times 8$ (6)

994. $(5 \times \square) - (\square \times 2) - 1 = \square \times 6$ (11)

995. $(4 \times \square) - (\square \times 9) + 1 = \square \times 5$ (3)

996. $(4 \times \square) - (\square \times 8) - 1 = \square \times 9$ (1)

997. $(4 \times \square) - (\square \times 7) + 1 = \square \times 6$ (4)

998. $(4 \times \square) - (\square \times 6) - 1 = (\square \times 7) + 1$ (2)

999. $(4 \times \square) - (\square \times 5) + 1 = \square \times 8$ (5)

1000. (4 X □) − (□ X 3) + 1 = (□ X 7) − 1   (8)

1001. (3 X □) − (□ X 9) + 1 = (□ X 5) − 1   (3)

1002. (3 X □) − (□ X 8) − 1 = □ X 9   (2)

1003. (3 X □) − (□ X 7) − 1 = (□ X 6) + 1   (3)

1004. (3 X □) − (□ X 6) − 1 = □ X 5   (4)

1005. (3 X □) − (□ X 5) − 1 = □ X 7   (2)

1006. (3 X □) − (□ X 4) + 1 = □ X 6   (6)

1007. (3 X □) − (□ X 2) − 1 = □ X 4   (13)

1008. (2 X □) − (□ X 8) − 1 = □ X 3   (2)

1009. (2 X □) − (□ X 7) + 1 = □ X 3   (3)

1010. (2 X □) − (□ X 6) + 1 = (□ X 7) − 1   (1)

1011. (2 X □) − (□ X 5) − 1 = □ X 3   (3)

1012. (2 X □) − (□ X 3) + 1 = □ X 4   (7)

1013. (2 X □) + (□ X 7) − 1 = □ X 9   (9)

1014. (2 X □) + (□ X 7) + 1 = □ X 8   (10)

1015. (2 X □) + (□ X 3) − 1 = □ X 4   (20)

Chapter Fifteen   A Variety Pack

Some of these problems are multiplication, while others are addition or subtraction. Just notice the signs and enjoy the challenge.

The first one (#1016) and the last one (#1055) have two solutions. The others have one.

1016.   3 □
       X ___□
         2 8 8

1017.   3 □
       X ___□
         1 9 8

1018.   3 □
       X ___□
         1 0 8

1019.   3 □
       X ___□
          7 8

1020.   3 □
       X ___□
         2 2 8

1021.   3 □
       X ___□
         2 3 8

1022.   3 □
       X ___□
         1 4 8

1023.   3 □
       X ___□
         1 2 8

1024.   4 □
       X ___□
         3 2 9

1025.   6 □
       X ___□
         5 4 4

1026.   □ □
       X ___4
         3 0 8

1027.   □ 4
       X ___□
         6 7 2

1028.   □ 6
       X ___□
         2 7 6

1029.   □ 4
       X ___□
         6 6 6

1030.   4 □
       X ___□
         4 1 4

1031.   □ 8
       X ___□
         2 6 6

1032.   3 □
       X ___□
         3 0 6

1033.   □ □
       X ___6
         5 8 8

1034.   8 □
       X ___□
         6 9 6

1035.   □ □
       X ___2
         1 7 8

1036.   □ 4
      + 3 □
         7 2

1037.   2 □
      + □ 5
         8 2

1038. □ 2 □
      + 5 □ 3
         9 2 1

1039. □ 4 □ 7 □
      + 1 □ 5 □ 3
        6 3 0 7 1

1040. □ 3 □
      - 4 □ 6
        2 3 9

1041. 8 0 4
      - □ □ □
        1 8 9

1042. □ 4 □ 0 2
      - 3 □ 6 □ 9 □
        5 4 5 0 9 8

1043. 8 □ 3
      - □ 7 □
        5 3 4

1044.
```
   □□□
-  3 0 8
───────
   4 9 9
```

1045.
```
   7 4 6
-  □□□
───────
   4 9 8
```

1046.
```
   □4□
-  3□2
───────
   4 5 9
```

1047.
```
   □5
+  6□
───────
   9 4
```

1048.
```
   □2□
+  4□4
───────
   7 1 3
```

1049.
```
   □□□
+  2 5 6
───────
   8 4 1
```

1050.
```
   6□8
+  □3□
───────
   9 3 4
```

1051.
```
   6□
X    □
───────
   5 0 4
```

1052.
```
   6□
X    □
───────
   4 0 8
```

1053.
```
   6□
X    □
───────
   2 0 4
```

1054.
```
   □2
X    □
───────
   5 5 8
```

1055.
```
     □□
X    □□
───────
   1 4 0
   7 0
───────
   8 4 0
```

Chapter Sixteen   An Advanced Level of Puzzles

These puzzles range from some having three unknowns or empty boxes to others having as many as eight unknowns. The answer section only lists three solutions for each puzzle in this section. Each puzzle may likely have many more solutions than that. Therefore, your solution may well be a valid one, yet not be shown as one of the solutions in the answer section.

1056. $(\Box \times 4) - (\Box \times 7) - (\Box \times 5) = 1$ 1057. $(\Box \times 8) - (\Box \times 9) = (\Box \times 6) + 1$

1058. $(\Box \times 7) + (\Box \times 5) = (\Box \times 6) - 1$ 1059. $(\Box \times 5) - (\Box \times 6) - (\Box \times 7) = 2$

1060. $(\Box \times 7) - (\Box \times 8) - (\Box \times 9) = 0$ 1061. $(\Box \times 6) - (\Box \times 5) - (\Box \times 7) = 0$

1062. $(\Box \times 9) + (\Box \times 7) - (\Box \times 8) = 0$ 1063. $(\Box \times 7) - (\Box \times 8) - (\Box \times 9) = 2$

1064. $(\Box \times 6) - (\Box \times 7) - (\Box \times 5) = 1$ 1065. $(\Box \times 5) + (\Box \times 8) + 1 = \Box \times 6$

1066. $(\Box \times 4) + (\Box \times 7) - (\Box \times 9) = 2$ 1067. $(\Box \times 7) - (\Box \times 9) - (\Box \times 8) = 1$

1068. $(\Box \times 8) - (\Box \times 9) - (\Box \times 7) = 2$ 1069. $(\Box \times 7) + (\Box \times 8) - (\Box \times 9) = 1$

1070. $(\Box \times 4) - (\Box \times 7) - (\Box \times 9) = 0$ 1071. $(\Box \times 6) - (\Box \times 7) - (\Box \times 9) = 0$

1072. $(\Box \times \Box) - 1 = \Box \times 7$ 1073. $(\Box \times 8) + (\Box \times 5) + (\Box \times 7) - 1 = \Box \times 9$

1074. $(\Box \times 5) + (\Box \times 4) + (\Box \times 7) + 1 = \Box \times 9$

1075. $(\Box \times 7) + (\Box \times 8) + (\Box \times 5) = \Box \times 9$

1076. $(\Box \times 7) - (\Box \times 6) - (\Box \times 5) = (\Box \times 8) - 1$

1077. $(\Box \times 9) - (\Box \times 5) - (\Box \times 4) + 1 = \Box \times 6$

1078. $(\Box \times 6) - (\Box \times 7) - (\Box \times 8) - 1 = \Box \times 5$

1079. $(\Box \times 7) - (\Box \times 8) - (\Box \times 9) - (\Box \times 4) = 1$

1080. $(\Box \times 8) + (\Box \times 9) + (\Box \times 5) = \Box \times 6$

1081. $(\Box \times 6) - (\Box \times 7) - 1 = (\Box \times 9) + (\Box \times 4)$

1082. $(\Box \times 8) + (\Box \times 3) = (\Box \times 6) - (\Box \times 5)$

1083. $(\square X 3) = (\square X 8) - (\square X 7) + (\square X 5)$

1084. $(\square X 8) - (\square X 9) - (\square X 7) = \square X 6$

1085. $(\square X 4) - (\square X 5) + (\square X 9) - (\square X 7) = 1$

1086. $(\square X 8) - (\square X 7) + (\square X 9) - (\square X 4) = 0$

1087. $(\square X 4) - (\square X 9) + (\square X 5) = \square X 8$

1088. $(\square X 5) - (\square X 8) + (\square X 7) + 1 = \square X 9$

1089. $(\square X 4) + (\square X 7) + (\square X 9) = \square X 8$

1090. $(\square X 7) - (\square X 9) - (\square X 8) + 1 = \square X 5$

1091. $(\square X 6) - (\square X 7) - (\square X 5) = \square X 4$

1092. $(\square X 4) - (\square X 9) + (\square X 7) = \square X 8$

1093. $(\square X 3) - (\square X 7) + (\square X 5) - (\square X 3) = 0$

1094. $(\square X 4) - (\square X 9) + (\square X 8) - (\square X 3) = 0$

1095. $(\square X 9) - (\square X 8) - (\square X 7) - (\square X 5) = 0$

1096. $(\square X 5) - (\square X 9) + (\square X 4) - (\square X 8) = 0$

1097. $(\square X 6) - (\square X 5) + (\square X 8) - (\square X 9) = 0$

1098. $(\square X 8) - (\square X 5) + (\square X 7) - (\square X 6) = 1$

1099. $(\square X 6) - (\square X 7) + (\square X 8) - (\square X 9) = 1$

1100. $(\square X 4) - (\square X 9) + (\square X 6) - (\square X 2) = 1$

1101. $(\square X 2) - (\square X 5) + (\square X 9) - (\square X 7) = 1$

1102. $(\square X 5) + (\square X 7) + (\square X 9) - 1 = \square X 8$

1103. $(\square X 3) + (\square X 4) + (\square X 5) = \square X 7$

1104. $(\square X 4) + (\square X 7) + (\square X 3) - (\square X 5) = 0$

1105. $(\square X 5) + (\square X 7) - (\square X 8) - (\square X 3) = 0$

1106. $(\square \, X \, 5) - (\square \, X \, 8) + (\square \, X \, 7) + 1 = \square \, X \, 9$

1107. $(\square \, X \, 8) + (\square \, X \, 7) + (\square \, X \, 5) - (\square \, X \, 9) = 0$

1108. $(\square \, X \, 3) + (\square \, X \, 8) + (\square \, X \, 5) = \square \, X \, 6$

1109. $(\square \, X \, 5) - (\square \, X \, 7) + (\square \, X \, 4) - (\square \, X \, 9) = 0$

1110. $(\square \, X \, 3) - (\square \, X \, 7) + (\square \, X \, 8) + 1 = \square \, X \, 5$

1111. $(\square \, X \, 4) + (\square \, X \, 7) - (\square \, X \, 9) - 1 = \square \, X \, 6$

1112. $(\square \, X \, 7) - (\square \, X \, 9) + (\square \, X \, 3) = \square \, X \, 8$

1113. $(\square \, X \, 3) - (\square \, X \, 5) + (\square \, X \, 7) - (\square \, X \, 8) = 0$

1114. $(\square \, X \, 3) + (\square \, X \, 8) + (\square \, X \, 5) = \square \, X \, 6$

1115. $(\square \, X \, 6) + (\square \, X \, 7) + (\square \, X \, 5) = \square \, X \, 8$

1116. $(\square \, X \, 8) - (\square \, X \, 9) - (\square \, X \, 7) = \square \, X \, 6$

1117. $(\square \, X \, 4) + (\square \, X \, 7) + (\square \, X \, 9) = \square \, X \, 8$

1118. $(\square \, X \, 6) + (\square \, X \, 7) - (\square \, X \, 8) - (\square \, X \, 5) = 0$

1119. $(\square \, X \, 4) - (\square \, X \, 9) + (\square \, X \, 7) + 1 = \square \, X \, 6$

1120. $(\square \, X \, 7) - (\square \, X \, 9) - (\square \, X \, 8) = \square \, X \, 3$

1121. $(\square \, X \, 5) - (\square \, X \, 8) - (\square \, X \, 7) = \square \, X \, 3$

1122. $(\square \, X \, 4) + (\square \, X \, 5) - (\square \, X \, 8) - (\square \, X \, 3) = 0$

1123. $(\square \, X \, 5) - (\square \, X \, 7) - (\square \, X \, 9) - (\square \, X \, 8) = 0$

1124. $(\square \, X \, 7) - (\square \, X \, 8) - (\square \, X \, 9) - (\square \, X \, 4) = 0$

1125. $(\square \, X \, 6) - (\square \, X \, 7) - (\square \, X \, 5) - (\square \, X \, 3) = 0$

1126. $(\square \, X \, 9) - (\square \, X \, 7) + (\square \, X \, 5) - 1 = \square \, X \, 8$

1127. $(\square \, X \, 8) - (\square \, X \, 7) - (\square \, X \, 9) - (\square \, X \, 5) = 0$

1128. $(\square \, X \, 3) - (\square \, X \, 8) + (\square \, X \, 5) - (\square \, X \, 9) = 2$

1129. $(\square X 4) - (\square X 9) + (\square X 7) - (\square X 8) = 0$

1130. $(\square X 4) + (\square X 7) + (\square X 9) + 2 = \square X 8$

1131. $(\square X 4) - (\square X 7) - (\square X 8) + 1 = \square X 5$

1132. $(\square X 3) - (\square X 7) - (\square X 5) + 1 = \square X 6$

1133. $(\square X 4) + (\square X 7) + (\square X 9) - 2 = \square X 8$

1134. $(\square X 6) - (\square X 7) - (\square X 5) = \square X 2$

1135. $(\square X 3) + (\square X 7) + (\square X 5) + 2 = (\square X 9) - 1$

1136. $(\square X 4) + (\square X 7) + (\square X 5) + 1 = \square X 8$

1137. $(\square X 6) + (\square X 7) + (\square X 5) - 1 = \square X 8$

1138. $(\square X 7) - (\square X 8) - (\square X 9) - (\square X 6) = 0$

1139. $(\square X 6) + (\square X 5) + (\square X 7) = \square X 8$

1140. $(\square X 4) + (\square X 7) + (\square X 9) = \square X 8$

1141. $(\square X 9) + (\square X 8) + (\square X 7) - 1 = \square X 6$

1142. $(\square X 6) - (\square X 9) + (\square X 5) - (\square X 7) = 0$

1143. $(\square X 2) - (\square X 5) + (\square X 7) - (\square X 9) = 0$

1144. $(\square X 6) - (\square X 7) - (\square X 9) - (\square X 4) = 0$

1145. $(\square X 5) - (\square X 7) + (\square X 3) - 1 = \square X 4$

1146. $(\square X 7) - (\square X 8) + (\square X 9) - 1 = \square X 6$

1147. $(\square X 8) - (\square X 9) - (\square X 7) = \square X 6$

1148. $(\square X 5) - (\square X 9) + (\square X 7) + 1 = \square X 3$

1149. $(\square X 4) + (\square X 5) + (\square X 7) - 1 = \square X 6$

1150. $(\square X 6) - (\square X 5) + (\square X 7) + 1 = \square X 4$

1151. $(\square X 8) - (\square X 9) - (\square X 7) + 1 = \square X 6$

1152. $(\Box \times 6) - (\Box \times 7) - (\Box \times 8) = \Box \times 5$

1153. $(\Box \times 7) - (\Box \times 8) - (\Box \times 9) - 1 = \Box \times 6$

1154. $(\Box \times 4) + (\Box \times 7) - (\Box \times 9) = \Box \times 3$

1155. $(\Box \times 6) - (\Box \times 7) + (\Box \times 5) + 1 = \Box \times 9$

1156. $(\Box \times 4) - (\Box \times 9) + (\Box \times 3) + 1 = \Box \times 6$

1157. $(\Box \times 5) - (\Box \times 6) - (\Box \times 7) + 1 = \Box \times 3$

1158. $(\Box \times 5) - (\Box \times 8) - (\Box \times 9) + 1 = \Box \times 7$

1159. $(\Box \times 8) - (\Box \times 9) + (\Box \times 4) - 1 = \Box \times 5$

1160. $(\Box \times 5) + (\Box \times 7) + (\Box \times 8) + 1 = \Box \times 6$

1161. $(\Box \times 6) - (\Box \times 7) + (\Box \times 9) = \Box \times 5$

1162. $(\Box \times 7) - (\Box \times 5) - (\Box \times 9) = \Box \times 8$

1163. $(\Box \times 8) - (\Box \times 9) + (\Box \times 7) = \Box \times 6$

1164. $(\Box \times \Box) - 1 = \Box \times \Box$

1165. $(\Box \times \Box) - 1 = (\Box \times \Box) + 1$

1166. $(\Box \times 6) - (\Box \times 7) - (\Box \times 5) + 1 = \Box \times 9$

1167. $(\Box \times 4) + (\Box \times 7) - (\Box \times 9) - 1 = \Box \times 6$

1168. $(\Box \times 6) - (\Box \times 7) - (\Box \times 5) = \Box \times 8$

1169. $(\Box \times 7) - (\Box \times 8) - (\Box \times 5) = \Box \times 6$

1170. $(\Box \times 5) - (\Box \times 7) - (\Box \times 8) + 1 = \Box \times 6$

1171. $(\Box \times 4) - (\Box \times 9) + (\Box \times 7) - 1 = \Box \times 8$

1172. $(\Box \times 3) - (\Box \times 7) + (\Box \times 9) - 1 = \Box \times 6$

1173. $(\Box \times 3) - (\Box \times 7) + (\Box \times 8) + 1 = \Box \times 9$

1174. $(\Box \times 7) - (\Box \times 8) - (\Box \times 9) = \Box \times 5$

1175. $(\square X 5) - (\square X 8) - (\square X 9) = \square X 7$

1176. $(\square X 5) - (\square X 6) - (\square X 7) - (\square X 8) = 0$

1177. $\square X 5 = (\square X 9) + (\square X 4) + (\square X 6) + 1$

1178. $\square X 3 = (\square X 8) - (\square X 7) + (\square X 5)$

1179. $\square X 4 = (\square X 7) - (\square X 9) - (\square X 5)$

1180. $(\square X 7) - (\square X 8) = (\square X 5) - (\square X 9)$

1181. $(\square X 5) + (\square X 7) = (\square X 8) - (\square X 3)$

1182. $(\square X 5) - (\square X 6) + (\square X 8) - (\square X 7) = 0$

1183. $(\square X 3) - (\square X 7) + (\square X 8) - (\square X 9) = 0$

1184. $(\square X 6) - (\square X 7) + (\square X 5) - (\square X 9) = 0$

1185. $(\square X 4) - (\square X 5) + (\square X 7) - (\square X 9) = 0$

1186. $(\square X 7) - (\square X 8) - (\square X 9) - 1 = \square X 4$

1187. $(\square X 5) - (\square X 8) + (\square X 7) - (\square X 6) = 0$

1188. $\square X 7 = (\square X 9) + (\square X 8) - (\square X 5) - 1$

1189. $(\square X 8) - (\square X 9) - (\square X 7) = \square X 6$

1190. $(\square X 6) - (\square X 7) - (\square X 5) - (\square X 4) = 0$

1191. $(\square X 2) + (\square X 3) + (\square X 5) + (\square X 7) = \square X 6$

1192. $(\square X 8) - (\square X 7) - (\square X 9) - (\square X 5) = \square X 3$

1193. $(\square X 4) - (\square X 7) - (\square X 5) = (\square X 9) - (\square X 8) - 1$

1194. $(\square X 7) + (\square X 8) + (\square X 9) + (\square X 5) = \square X 6$

1195. $(\square X 5) + (\square X 4) + (\square X 3) + (\square X 7) = \square X 9$

1196. $(\square X 5) + (\square X 6) + (\square X 7) - 1 = (\square X 9) - (\square X 8)$

1197. $(\square X 5) - (\square X 7) - (\square X 9) = (\square X 8) - (\square X 6) - 1$

1198. $(\square X 4)-(\square X 9)-(\square X 5)=(\square X 7)-(\square X 8)$

1199. $(\square X 5)-(\square X 9)-(\square X 4)=(\square X 7)-(\square X 8)+1$

1200. $(\square X 5)+(\square X 4)+(\square X 7)=(\square X 8)-(\square X 3)+1$

1201. $(\square X 3)+(\square X 4)+(\square X 7)-(\square X 8)-(\square X 5)=0$

1202. $(\square X 5)-(\square X 9)-(\square X 6)+(\square X 7)-1=\square X 3$

1203. $(\square X 4)-(\square X 9)-(\square X 5)+(\square X 7)-(\square X 8)=1$

1204. $(\square X 5)-(\square X 9)-(\square X 7)+(\square X 8)-(\square X 3)=0$

1205. $(\square X 8)-(\square X 5)-(\square X 7)+(\square X 9)+1=\square X 4$

1206. $(\square X 6)+(\square X 7)+(\square X 5)+(\square X 2)-(\square X 9)=0$

1207. $(\square X 8)-(\square X 9)-(\square X 7)+(\square X 5)=\square X 2$

1208. $(\square X 6)-(\square X 7)-(\square X 5)+(\square X 9)-(\square X 4)=0$

1209. $(\square X 5)-(\square X 9)-(\square X 6)+(\square X 7)-(\square X 3)=0$

1210. $(\square X 7)-(\square X 8)-(\square X 9)-(\square X 5)=\square X 3$

1211. $(\square X 6)-(\square X 7)+(\square X 9)-(\square X 8)=\square X 3$

1212. $(\square X 3)-(\square X 7)=(\square X 8)+(\square X 9)-(\square X 6)$

1213. $(\square X 3)-(\square X 7)+(\square X 8)-(\square X 5)=\square X 4$

1214. $(\square X 7)-(\square X 8)+(\square X 6)-(\square X 9)=\square X 5$

1215. $(\square X 4)-(\square X 9)-(\square X 7)+(\square X 5)=\square X 6$

1216. $(\square X 5)+(\square X 6)+(\square X 7)-(\square X 8)-(\square X 3)=0$

1217. $(\square X 4)+(\square X 5)+(\square X 7)+(\square X 3)=\square X 9$

1218. $(\square X 5)-(\square X 9)-(\square X 7)+(\square X 6)-(\square X 3)=0$

1219. $(\square X 7)-(\square X 8)-(\square X 9)+(\square X 5)=\square X 2$

1220. $(\square X 4)+(\square X 9)+(\square X 5)+(\square X 8)=\square X 6$

1221. $(\square \times 3) + (\square \times 8) - (\square \times 7) - (\square \times 5) = \square \times 4$

1222. $(\square \times 8) - (\square \times 9) - (\square \times 7) - (\square \times 5) = \square \times 3$

1223. $(\square \times 3) - (\square \times 8) + (\square \times 4) + (\square \times 9) = \square \times 6$

1224. $(\square \times 6) - (\square \times 8) + (\square \times 5) - (\square \times 9) = \square \times 4$

1225. $(\square \times 4) - (\square \times 9) - (\square \times 5) + (\square \times 3) - (\square \times 6) = 0$

1226. $(\square \times 3) - (\square \times 7) - (\square \times 8) + (\square \times 5) - (\square \times 6) = 0$

1227. $(\square \times 6) - (\square \times 7) + (\square \times 5) + (\square \times 3) = \square \times 8$

1228. $(\square \times 7) - (\square \times 9) - (\square \times 8) + (\square \times 6) + 1 = \square \times 7$

1229. $(\square \times 5) + (\square \times 6) + (\square \times 7) + (\square \times 4) + 1 = \square \times 8$

1230. $(\square \times 4) - (\square \times 7) - (\square \times 3) + (\square \times 8) - (\square \times 6) = 0$

1231. $(\square \times 4) - (\square \times 5) - (\square \times 7) + (\square \times 8) - (\square \times 9) = 0$

1232. $(\square \times 4) + (\square \times 5) + (\square \times 3) + (\square \times 7) + 1 = \square \times 8$

1233. $(\square \times 8) + (\square \times 9) + (\square \times 7) - 1 = (\square \times 6) + (\square \times 5)$

1234. $(\square \times 4) + (\square \times 9) + (\square \times 7) + (\square \times 5) - 1 = (\square \times 8) + 1$

1235. $(\square \times 3) + (\square \times 8) + (\square \times 7) + (\square \times 5) + 1 = (\square \times 9) - 1$

1236. $(\square \times 7) - (\square \times 8) - (\square \times 9) + (\square \times 5) - 1 = \square \times 6$

1237. $(\square \times 6) - (\square \times 7) - (\square \times 5) + (\square \times 9) + 1 = \square \times 8$

1238. $(\square \times 5) - (\square \times 7) - (\square \times 9) + (\square \times 8) - (\square \times 6) = 1$

1239. $(\square \times 4) - (\square \times 7) - (\square \times 5) + (\square \times 9) + 1 = \square \times 8$

1240. $(\square \times 6) - (\square \times 7) + (\square \times 5) + (\square \times 3) = \square \times 8$

1241. $(\square \times 4) - (\square \times 5) - (\square \times 3) + (\square \times 8) = \square \times 9$

1242. $(\square \times 2) - (\square \times 3) - (\square \times 5) + (\square \times 9) = \square \times 6$

1243. $(\square \times 5) - (\square \times 7) - (\square \times 8) + (\square \times 9) + 1 = \square \times 4$

1244. $(\square \times 2) - (\square \times 9) - (\square \times 5) + (\square \times 8) + 1 = \square \times 3$

1245. $(\square \times 3) - (\square \times 7) - (\square \times 4) + (\square \times 9) + 1 = \square \times 5$

1246. $(\square \times 6) - (\square \times 4) - (\square \times 7) + (\square \times 9) = \square \times 8$

1247. $(\square \times 4) - (\square \times 5) - (\square \times 7) + (\square \times 8) - 1 = \square \times 6$

1248. $(\square \times 4) - (\square \times 9) - (\square \times 3) + (\square \times 7) - 1 = \square \times 5$

1249. $(\square \times 7) + (\square \times 2) - (\square \times 9) - (\square \times 8) = \square \times 4$

1250. $(\square \times 7) - (\square \times 9) - (\square \times 8) + (\square \times 5) + 1 = \square \times 6$

1251. $(\square \times 4) - (\square \times 9) + (\square \times 5) - (\square \times 7) - 1 = \square \times 3$

1252. $(\square \times 6) - (\square \times 7) - (\square \times 5) + (\square \times 8) = \square \times 9$

1253. $(\square \times 8) - (\square \times 9) - (\square \times 7) - (\square \times 5) = \square \times 3$

1254. $(\square \times 3) - (\square \times 8) - (\square \times 7) + (\square \times 9) - (\square \times 5) = 0$

1255. $(\square \times 2) - (\square \times 7) + (\square \times 9) - (\square \times 8) - (\square \times 5) = 0$

1256. $\square \times 8 = (\square \times 7) - (\square \times 5) - (\square \times 3) - (\square \times 2)$

1257. $\square \times 5 = (\square \times 6) + (\square \times 7) + (\square \times 8) - (\square \times 9)$

1258. $(\square \times 4) - (\square \times 7) + (\square \times 9) - (\square \times 5) = \square \times 8$

1259. $(\square \times 3) - (\square \times 8) + (\square \times 7) + (\square \times 5) - (\square \times 6) = 0$

1260. $(\square \times 5) - (\square \times 7) - (\square \times 9) + (\square \times 8) = \square \times 3$

1261. $(\square \times 3) - (\square \times 7) - (\square \times 8) + (\square \times 9) = \square \times 4$

1262. $(\square \times 8) - (\square \times 7) - (\square \times 9) - (\square \times 5) = \square \times 2$

1263. $(\square \times 6) - (\square \times 8) + (\square \times 9) + (\square \times 5) = \square \times 3$

1264. $(\square \times 9) - (\square \times 8) - (\square \times 5) + (\square \times 6) + 1 = \square \times 4$

1265. $\square \times 6 = (\square \times 7) + (\square \times 5) + (\square \times 4) + (\square \times 9)$

1266. $(\square \times 7) - (\square \times 8) + (\square \times 9) - (\square \times 5) - (\square \times 6) = 0$

1267. $(\square \times 3) + (\square \times 7) - (\square \times 5) + (\square \times 8) - (\square \times 9) = 0$

1268. $(\square \times 3) + (\square \times 7) - (\square \times 5) + (\square \times 8) - (\square \times 9) = 1$

1269. $(\square \times 4) + (\square \times 7) + (\square \times 9) - (\square \times 5) - (\square \times 3) = 0$

1270. $(\square \times 4) - (\square \times 5) - (\square \times 9) + (\square \times 7) + (\square \times 8) - 1 = \square \times 5$

1271. $(\square \times 7) - (\square \times 8) - (\square \times 3) + (\square \times 9) + (\square \times 5) - (\square \times 6) = 0$

1272. $(\square \times 4) - (\square \times 5) - (\square \times 7) + (\square \times 8) - (\square \times 9) - (\square \times 2) = 0$

1273. $(\square \times 9) - (\square \times 8) - (\square \times 7) + (\square \times 6) + (\square \times 5) - 1 = \square \times 8$

1274. $(\square \times 6) + (\square \times 7) + (\square \times 8) + (\square \times 9) + (\square \times 4) - 1 = \square \times 7$

1275. $(\square \times 7) + (\square \times 8) + (\square \times 9) - (\square \times 6) = (\square \times 5) + (\square \times 3)$

1276. $(\square \times 8) - (\square \times 9) - (\square \times 7) = (\square \times 4) + (\square \times 5) - (\square \times 2)$

1277. $(\square \times 8) - (\square \times 7) - (\square \times 9) = (\square \times 6) + (\square \times 5) - (\square \times 4)$

1278. $(\square \times 4) - (\square \times 7) - (\square \times 5) = (\square \times 8) - (\square \times 9) + (\square \times 2)$

1279. $(\square \times 2) - (\square \times 3) - (\square \times 5) + 1 = (\square \times 9) + (\square \times 4) - (\square \times 7) - 1$

1280. $(\square \times 6) - (\square \times 7) - (\square \times 5) = (\square \times 8) - (\square \times 9) - (\square \times 3)$

1281. $(\square \times 7) - (\square \times 8) - (\square \times 5) = (\square \times 9) - (\square \times 4) - (\square \times 6)$

1282. $(\square \times 4) - (\square \times 9) - (\square \times 7) = (\square \times 8) - (\square \times 5) - (\square \times 6) - 1$

1283. $(\square \times 6) + (\square \times 7) + (\square \times 5) - (\square \times 9) - (\square \times 8) = \square \times 3$

1284. $(\square \times 7) - (\square \times 9) - (\square \times 8) + (\square \times 5) - (\square \times 6) = \square \times 4$

1285. $(\square \times 8) - (\square \times 7) - (\square \times 9) + (\square \times 5) - (\square \times 6) = \square \times 3$

1286. $(\square \times 2) + (\square \times 3) + (\square \times 5) + (\square \times 9) = (\square \times 7) + (\square \times 8)$

1287. $(\square \times 3) + (\square \times 8) + (\square \times 7) + (\square \times 5) + 1 = (\square \times 9) - (\square \times 2)$

1288. $(\square \times 5) - (\square \times 7) - (\square \times 4) + (\square \times 9) = (\square \times 8) - (\square \times 9) - 1$

1289. $(\square \times 3) - (\square \times 8) + (\square \times 7) - (\square \times 9) + (\square \times 6) - (\square \times 2) = 0$

1290. $(\square X 9) - (\square X 8) - (\square X 7) + (\square X 5) + (\square X 3) = \square X 6$

1291. $(\square X 5) + (\square X 3) + (\square X 9) - (\square X 7) + (\square X 8) = \square X 4$

1292. $(\square X 3) - (\square X 8) - (\square X 5) + (\square X 7) + (\square X 2) = \square X 6$

1293. $(\square X 9) - (\square X 8) - (\square X 7) - (\square X 4) = (\square X 6) - (\square X 5) - 1$

1294. $\square X 9 = (\square X 8) - (\square X 7) - (\square X 5) - (\square X 4) + (\square X 3)$

1295. $(\square X 7) - (\square X 9) - (\square X 8) + (\square X 5) - (\square X 4) = \square X 3$

1296. $(\square X 7) + (\square X 8) - (\square X 9) + (\square X 5) + (\square X 6) = \square X 4$

1297. $(\square X 3) - (\square X 7) + (\square X 5) - (\square X 8) + (\square X 9) = \square X 4$

1298. $(\square X 4) - (\square X 7) - (\square X 9) + (\square X 5) + (\square X 3) = \square X 8$

1299. $(\square X 7) - (\square X 8) - (\square X 9) + (\square X 5) = (\square X 6) - (\square X 2)$

1300. $(\square X 2) + (\square X 3) + (\square X 5) - (\square X 9) - (\square X 7) = \square X 8$

1301. $(\square X 7) + (\square X 8) + (\square X 9) = (\square X 6) + (\square X 5) - (\square X 3) - 1$

1302. $(\square X 6) - (\square X 7) + (\square X 5) - (\square X 9) - 1 = (\square X 8) + (\square X 3) - (\square X 7)$

1303. $(\square X 6) - (\square X 7) - (\square X 5) - (\square X 9) = (\square X 8) - (\square X 3) + (\square X 5)$

1304. $(\square X 3) - (\square X 4) - (\square X 5) = (\square X 8) - (\square X 9) - (\square X 6) + (\square X 3)$

1305. $(\square X 6) - (\square X 7) - (\square X 5) + (\square X 8) + (\square X 9) = (\square X 3) + (\square X 5) + 1$

1306. $(\square X 3) + (\square X 4) + (\square X 7) - (\square X 9) - (\square X 8) = (\square X 6) - (\square X 5)$

1307. $(\square X 5) - (\square X 3) - (\square X 7) + (\square X 9) - (\square X 8) = (\square X 6) - (\square X 5)$

1308. $(\square X 7) - (\square X 8) - (\square X 9) + (\square X 5) - (\square X 6) = (\square X 9) - (\square X 7)$

1309. $(\square X 5) + (\square X 7) + (\square X 9) + (\square X 2) = (\square X 6) + (\square X 8) + (\square X 3) - 1$

1310. $(\square X 3) - (\square X 8) + (\square X 7) - (\square X 9) + (\square X 5) = (\square X 6) - (\square X 4)$

1311. $(\square X 3) + (\square X 4) + (\square X 5) + (\square X 7) - (\square X 9) = (\square X 6) - (\square X 5)$

1312. $(\square X 8) + (\square X 9) + (\square X 5) - (\square X 7) - 1 = (\square X 2) + (\square X 3) - (\square X 4)$

1313. $(\square X 5) + (\square X 6) + (\square X 7) = (\square X 9) - (\square X 8) - (\square X 3) + (\square X 4)$

Because the next several puzzles are so long, each one is being shown using two lines. The break-up will be after the equal sign.

1314. $(\square X 7) - (\square X 9) - (\square X 5) + (\square X 8) + (\square X 3) =$

$(\square X 6) - (\square X 7) + (\square X 4)$

1315. $(\square X 9) - (\square X 5) - (\square X 4) + (\square X 7) =$

$(\square X 3) + (\square X 8) - (\square X 7) - (\square X 5)$

1316. $(\square X 2) + (\square X 3) + (\square X 7) - (\square X 8) =$

$(\square X 5) - (\square X 9) + (\square X 6) - (\square X 4) - 1$

1317. $(\square X 5) + (\square X 8) + (\square X 3) - (\square X 7) =$

$(\square X 9) - (\square X 4) + (\square X 3) - (\square X 7)$

1318. $(\square X 8) - (\square X 7) - (\square X 5) + (\square X 9) - (\square X 4) =$

$(\square X 6) + (\square X 5) - (\square X 3)$

1319. $(\square X 5) + (\square X 7) - (\square X 4) + (\square X 9) - (\square X 6) =$

$(\square X 8) + (\square X 3) - (\square X 2)$

1320. $(\square X 4) + (\square X 5) - (\square X 7) - (\square X 3) =$

$(\square X 8) + (\square X 9) - (\square X 5) - (\square X 4)$

Chapter Seventeen   The Most Challenging Puzzles

You now have a chance to truly prove yourself. Your mind is the sharpest it's been in some time, and solving these next two puzzles will show just what a great mind you have developed. Be patient and persistent, as each of these puzzles could take an hour or more to find all possible solutions. The first one has forty-four solutions. The challenge comes in developing a strategy to use to find all the solutions. You'll need paper, but probably not a calculator. As you'll see in the answer section, I discovered two additional answers the second time around.

Try the first one now, and we'll discuss the second puzzle later. Good luck!

$$1321. \quad (\,9 \times \square\,) - (\,\square \times 2\,) + 1 = (\,\square \times 7\,) + (\,\square \times 8\,) - 1$$

You don't have to find all the solutions to that puzzle in one sitting. Take a break and try again, later.

With this next puzzle, see how many solutions you can find first without finding out how many solutions are in the answer section for this puzzle. When you've exhausted all possibilities and think you have them all, then check and see how your number of solutions compares with what's in the solution section. Here it is!

$$1322. \quad (\,4 \times \square\,) + (\,\square \times 7\,) + 1 = (\,\square \times 8\,) - (\,\square \times 3\,) - 1$$

Chapter Eighteen   The Help Section

A. Helpful Hints

When I was a student in elementary school back in the 1950's, my classmates and I had to memorize the multiplication tables – know them by heart – so that 7 X 6 = 42 was stated like one word with many syllables. That's how quickly the 42 was expected to be said. Perhaps you had to memorize them in the same way. However, if you haven't learned the tables yet, that's not a problem. You can learn them now. Just don't try to learn them as some students tend to do and that is to go up the ladder to get to the 7 X 6. In other words, don't try finding the solution by saying 7 X 1 = 7 …   7 X 2 = 14 … 7 X 3 = 21 … and eventually 7 X 6 = 42. That method takes too long which may have a negative effect in that you get bored because it takes so long and you lose interest and quit. Just memorize the answer immediately following the problem.

What's a good way to learn the multiplication tables? Don't try to overwhelm yourself by trying to memorize all of them in one sitting. Instead, allow yourself one day to memorize the "ones". That's 1 X 0 = 0 … 1 X 1 = 1 … 1 X 2 = 2 … and so on until you get to 1 X 9 = 9. That's all, just 0 – 9. But learn them backwards and forwards. Learn both 1 X 8 = 8 and 8 X 1 = 8. So in one day, you master the ones, whether it takes you thirty minutes or two hours. On the second day, learn the 2's, and on the third day, the 3's. Then stop and test yourself using the test on page 91 that covers only the 0's, 1's, 2's and 3's. Yes, go ahead and memorize the 0's, also, if you don't already know them, even though 0's are not used in this book's puzzles.

The test is a timed test. You must time yourself, because you need to complete the test in three minutes and twenty seconds. That's 2 seconds per problem for the first 100 problems, and then you have ten bonus problems for the truly fast ones. Hopefully, all of you will be able to complete all 110 problems in each test in three minutes and twenty seconds or less while getting every problem right. If you missed some, then study some more and take the test again.

When you've mastered the 0's, 1's, 2's and 3's, then take a day for learning the fours, another day for the fives, and a day for the sixes. Then stop and take another test – the 4, 5 and 6 timed test. Master that material and proceed with the 7, 8, and 9 during the next three days followed by your taking the 7, 8, and 9 test. There is a fourth test that covers all the multiplication facts from zero through nine. If you can get every answer right on that test within the three minutes and twenty seconds allowed, you can congratulate yourself, as you are now ready for the puzzles without a need for any crutches, such as needing to have the written tables in front of you. Those tables will be in your head ready to spin through the possibilities for solving each puzzle.

The next four pages will have the four timed tests. You may wish to get several copies made of each, as you may need them if you initially don't get them all right in the time allotted. Go ahead and take a look at them to see what's ahead of you. I think you'll recall a lot more than you may think, and maybe you won't need additional study.

B.   **0 – 3 Multiplication Tables' Test**

(Time yourself for 3 minutes and 20 seconds)

1) 5 x 1 = ____	23) 4 x 2 = ____	45) 1 x 3 = ____	67) 3 x 7 = ____	89) 3 x 1 = ____
2) 6 x 3 = ____	24) 6 x 0 = ____	46) 8 x 3 = ____	68) 1 x 3 = ____	90) 3 x 4 = ____
3) 0 x 4 = ____	25) 5 x 2 = ____	47) 6 x 1 = ____	69) 8 x 1 = ____	91) 2 x 8 = ____
4) 8 x 3 = ____	26) 6 x 3 = ____	48) 6 x 2 = ____	70) 5 x 2 = ____	92) 3 x 0 = ____
5) 0 x 3 = ____	27) 8 x 2 = ____	49) 9 x 0 = ____	71) 6 x 3 = ____	93) 2 x 4 = ____
6) 1 x 6 = ____	28) 6 x 2 = ____	50) 8 x 1 = ____	72) 9 x 1 = ____	94) 2 x 1 = ____
7) 0 x 5 = ____	29) 0 x 7 = ____	51) 5 x 3 = ____	73) 7 x 0 = ____	95) 9 x 2 = ____
8) 3 x 3 = ____	30) 1 x 8 = ____	52) 9 x 2 = ____	74) 3 x 2 = ____	96) 7 x 3 = ____
9) 1 x 2 = ____	31) 5 x 3 = ____	53) 7 x 2 = ____	75) 9 x 1 = ____	97) 4 x 2 = ____
10) 2 x 9 = ____	32) 4 x 2 = ____	54) 8 x 3 = ____	76) 7 x 3 = ____	98) 2 x 3 = ____
11) 4 x 3 = ____	33) 0 x 1 = ____	55) 0 x 6 = ____	77) 9 x 3 = ____	99) 9 x 0 = ____
12) 3 x 8 = ____	34) 1 x 4 = ____	56) 2 x 5 = ____	78) 8 x 3 = ____	100) 1 x 2 = ____
13) 3 x 5 = ____	35) 1 x 1 = ____	57) 7 x 2 = ____	79) 7 x 1 = ____	101) 5 x 3 = ____
14) 2 x 3 = ____	36) 0 x 0 = ____	58) 1 x 7 = ____	80) 3 x 9 = ____	102) 8 x 2 = ____
15) 8 x 3 = ____	37) 1 x 8 = ____	59) 0 x 2 = ____	81) 7 x 3 = ____	103) 6 x 3 = ____
16) 6 x 2 = ____	38) 3 x 9 = ____	60) 1 x 5 = ____	82) 8 x 2 = ____	104) 2 x 9 = ____
17) 3 x 3 = ____	39) 2 x 6 = ____	61) 0 x 9 = ____	83) 9 x 3 = ____	105) 3 x 8 = ____
18) 8 x 2 = ____	40) 3 x 2 = ____	62) 2 x 0 = ____	84) 0 x 8 = ____	106) 7 x 2 = ____
19) 0 x 3 = ____	41) 1 x 0 = ____	63) 2 x 7 = ____	85) 9 x 2 = ____	107) 9 x 3 = ____
20) 2 x 2 = ____	42) 7 x 2 = ____	64) 3 x 6 = ____	86) 7 x 3 = ____	108) 3 x 6 = ____
21) 4 x 3 = ____	43) 9 x 3 = ____	65) 7 x 2 = ____	87) 1 x 9 = ____	109) 9 x 2 = ____
22) 5 x 1 = ____	44) 7 x 1 = ____	66) 6 x 3 = ____	88) 2 x 2 = ____	110) 3 x 3 = ____

C.   **4 – 6 Multiplication Tables' Test**

(Time yourself for 3 minutes and 20 seconds)

1) 4 x 5 = ____	23) 5 x 5 = ____	45) 6 x 2 = ____	67) 9 x 4 = ____	89) 6 x 7 = ____
2) 3 x 6 = ____	24) 6 x 3 = ____	46) 6 x 7 = ____	68) 9 x 6 = ____	90) 4 x 6 = ____
3) 7 x 4 = ____	25) 9 x 6 = ____	47) 4 x 8 = ____	69) 6 x 2 = ____	91) 6 x 3 = ____
4) 4 x 9 = ____	26) 9 x 5 = ____	48) 6 x 0 = ____	70) 4 x 9 = ____	92) 5 x 4 = ____
5) 5 x 8 = ____	27) 7 x 6 = ____	49) 9 x 5 = ____	71) 6 x 9 = ____	93) 6 x 6 = ____
6) 6 x 7 = ____	28) 6 x 4 = ____	50) 4 x 7 = ____	72) 5 x 5 = ____	94) 3 x 4 = ____
7) 9 x 6 = ____	29) 6 x 8 = ____	51) 6 x 4 = ____	73) 8 x 4 = ____	95) 4 x 9 = ____
8) 1 x 5 = ____	30) 6 x 5 = ____	52) 7 x 5 = ____	74) 6 x 7 = ____	96) 6 x 1 = ____
9) 5 x 9 = ____	31) 9 x 4 = ____	53) 4 x 9 = ____	75) 8 x 6 = ____	97) 8 x 6 = ____
10) 6 x 8 = ____	32) 5 x 2 = ____	54) 4 x 4 = ____	76) 4 x 8 = ____	98) 9 x 4 = ____
11) 8 x 5 = ____	33) 3 x 5 = ____	55) 5 x 8 = ____	77) 8 x 5 = ____	99) 6 x 3 = ____
12) 4 x 6 = ____	34) 2 x 6 = ____	56) 1 x 6 = ____	78) 3 x 6 = ____	100) 8 x 6 = ____
13) 5 x 7 = ____	35) 6 x 3 = ____	57) 9 x 6 = ____	79) 6 x 8 = ____	101) 5 x 9 = ____
14) 4 x 8 = ____	36) 6 x 7 = ____	58) 4 x 7 = ____	80) 7 x 4 = ____	102) 3 x 6 = ____
15) 6 x 9 = ____	37) 4 x 6 = ____	59) 6 x 6 = ____	81) 5 x 8 = ____	103) 6 x 8 = ____
16) 5 x 6 = ____	38) 6 x 1 = ____	60) 5 x 9 = ____	82) 6 x 3 = ____	104) 2 x 6 = ____
17) 7 x 5 = ____	39) 7 x 4 = ____	61) 7 x 6 = ____	83) 4 x 7 = ____	105) 9 x 5 = ____
18) 8 x 6 = ____	40) 6 x 9 = ____	62) 8 x 4 = ____	84) 7 x 6 = ____	106) 6 x 4 = ____
19) 4 x 8 = ____	41) 7 x 6 = ____	63) 6 x 9 = ____	85) 6 x 6 = ____	107) 8 x 4 = ____
20) 6 x 6 = ____	42) 9 x 4 = ____	64) 6 x 8 = ____	86) 6 x 4 = ____	108) 6 x 9 = ____
21) 5 x 4 = ____	43) 5 x 7 = ____	65) 4 x 6 = ____	87) 7 x 5 = ____	109) 4 x 5 = ____
22) 8 x 4 = ____	44) 8 x 6 = ____	66) 9 x 6 = ____	88) 3 x 6 = ____	110) 8 x 6 = ____

D. 7 – 9 Multiplication Tables' Test

(Time yourself for 3 minutes and 20 seconds)

1) 7 x 9 = ____	23) 6 x 7 = ____	45) 4 x 9 = ____	67) 3 x 7 = ____	89) 8 x 6 = ____
2) 8 x 8 = ____	24) 7 x 3 = ____	46) 7 x 7 = ____	68) 9 x 9 = ____	90) 7 x 6 = ____
3) 4 x 7 = ____	25) 9 x 9 = ____	47) 9 x 6 = ____	69) 7 x 4 = ____	91) 9 x 4 = ____
4) 3 x 8 = ____	26) 7 x 6 = ____	48) 6 x 7 = ____	70) 7 x 9 = ____	92) 3 x 9 = ____
5) 9 x 7 = ____	27) 8 x 0 = ____	49) 3 x 7 = ____	71) 8 x 8 = ____	93) 8 x 7 = ____
6) 8 x 2 = ____	28) 6 x 9 = ____	50) 8 x 7 = ____	72) 3 x 8 = ____	94) 7 x 9 = ____
7) 8 x 9 = ____	29) 8 x 7 = ____	51) 9 x 8 = ____	73) 7 x 7 = ____	95) 6 x 8 = ____
8) 7 x 4 = ____	30) 5 x 7 = ____	52) 9 x 9 = ____	74) 2 x 9 = ____	96) 9 x 6 = ____
9) 8 x 5 = ____	31) 6 x 8 = ____	53) 7 x 6 = ____	75) 8 x 3 = ____	97) 8 x 7 = ____
10) 7 x 8 = ____	32) 7 x 5 = ____	54) 6 x 8 = ____	76) 8 x 7 = ____	98) 4 x 7 = ____
11) 8 x 6 = ____	33) 4 x 8 = ____	55) 8 x 8 = ____	77) 4 x 9 = ____	99) 3 x 8 = ____
12) 8 x 3 = ____	34) 8 x 9 = ____	56) 6 x 7 = ____	78) 7 x 9 = ____	100) 7 x 0 = ____
13) 9 x 6 = ____	35) 6 x 7 = ____	57) 3 x 9 = ____	79) 8 x 8 = ____	101) 6 x 9 = ____
14) 8 x 7 = ____	36) 7 x 8 = ____	58) 5 x 8 = ____	80) 9 x 4 = ____	102) 4 x 8 = ____
15) 9 x 4 = ____	37) 9 x 7 = ____	59) 4 x 7 = ____	81) 7 x 6 = ____	103) 3 x 9 = ____
16) 7 x 9 = ____	38) 6 x 9 = ____	60) 7 x 7 = ____	82) 4 x 7 = ____	104) 7 x 8 = ____
17) 9 x 3 = ____	39) 3 x 8 = ____	61) 3 x 9 = ____	83) 9 x 7 = ____	105) 9 x 8 = ____
18) 3 x 7 = ____	40) 7 x 8 = ____	62) 8 x 9 = ____	84) 8 x 4 = ____	106) 4 x 9 = ____
19) 8 x 4 = ____	41) 9 x 2 = ____	63) 9 x 7 = ____	85) 7 x 8 = ____	107) 8 x 4 = ____
20) 7 x 7 = ____	42) 4 x 8 = ____	64) 9 x 3 = ____	86) 6 x 9 = ____	108) 3 x 8 = ____
21) 8 x 1 = ____	43) 7 x 9 = ____	65) 4 x 9 = ____	87) 9 x 8 = ____	109) 5 x 9 = ____
22) 9 x 8 = ____	44) 6 x 9 = ____	66) 7 x 8 = ____	88) 8 x 4 = ____	110) 9 x 2 = ____

E. 0 – 9 Multiplication Tables' Test

(Time yourself for 3 minutes and 20 seconds)

1) 5 x 7 = ____	23) 0 x 7 = ____	45) 5 x 4 = ____	67) 8 x 1 = ____	89) 6 x 3 = ____
2) 1 x 4 = ____	24) 4 x 6 = ____	46) 3 x 2 = ____	68) 5 x 6 = ____	90) 2 x 5 = ____
3) 2 x 8 = ____	25) 5 x 1 = ____	47) 5 x 0 = ____	69) 9 x 2 = ____	91) 3 x 9 = ____
4) 5 x 3 = ____	26) 2 x 0 = ____	48) 9 x 8 = ____	70) 3 x 5 = ____	92) 8 x 6 = ____
5) 0 x 0 = ____	27) 9 x 7 = ____	49) 6 x 2 = ____	71) 6 x 1 = ____	93) 7 x 3 = ____
6) 3 x 1 = ____	28) 6 x 4 = ____	50) 8 x 5 = ____	72) 5 x 5 = ____	94) 0 x 6 = ____
7) 0 x 9 = ____	29) 1 x 0 = ____	51) 4 x 9 = ____	73) 0 x 4 = ____	95) 6 x 5 = ____
8) 3 x 8 = ____	30) 7 x 5 = ____	52) 9 x 6 = ____	74) 7 x 8 = ____	96) 9 x 4 = ____
9) 6 x 7 = ____	31) 4 x 8 = ____	53) 3 x 6 = ____	75) 3 x 3 = ____	97) 1 x 6 = ____
10) 2 x 4 = ____	32) 8 x 2 = ____	54) 5 x 2 = ____	76) 2 x 1 = ____	98) 7 x 2 = ____
11) 6 x 9 = ____	33) 7 x 6 = ____	55) 8 x 4 = ____	77) 8 x 7 = ____	99) 5 x 9 = ____
12) 1 x 5 = ____	34) 4 x 5 = ____	56) 4 x 3 = ____	78) 1 x 1 = ____	100) 8 x 8 = ____
13) 2 x 3 = ____	35) 1 x 3 = ____	57) 6 x 8 = ____	79) 9 x 6 = ____	101) 4 x 7 = ____
14) 8 x 3 = ____	36) 9 x 5 = ____	58) 9 x 9 = ____	80) 6 x 6 = ____	102) 8 x 7 = ____
15) 7 x 7 = ____	37) 1 x 7 = ____	59) 2 x 7 = ____	81) 3 x 8 = ____	103) 6 x 8 = ____
16) 3 x 9 = ____	38) 4 x 4 = ____	60) 3 x 4 = ____	82) 7 x 4 = ____	104) 9 x 9 = ____
17) 9 x 6 = ____	39) 8 x 9 = ____	61) 8 x 6 = ____	83) 9 x 8 = ____	105) 7 x 8 = ____
18) 7 x 3 = ____	40) 7 x 7 = ____	62) 3 x 7 = ____	84) 9 x 3 = ____	106) 9 x 7 = ____
19) 4 x 2 = ____	41) 4 x 7 = ____	63) 6 x 4 = ____	85) 7 x 8 = ____	107) 8 x 8 = ____
20) 7 x 9 = ____	42) 5 x 8 = ____	64) 9 x 1 = ____	86) 2 x 2 = ____	108) 6 x 9 = ____
21) 4 x 8 = ____	43) 2 x 6 = ____	65) 4 x 7 = ____	87) 4 x 9 = ____	109) 7 x 7 = ____
22) 2 x 9 = ____	44) 9 x 7 = ____	66) 3 x 9 = ____	88) 3 x 7 = ____	110) 4 x 9 = ____

F. Individualized Schedule for Mastering the Multiplication Tables

Okay, so you've discovered you didn't do well on the multiplication tests. That's not really a problem, because here you can have your own daily schedule for mastering the tables. All you have to do is decide how much time you can spend per day and then how many days you need to master each column. That's all. If you just try mastering one column per day, in ten days or less, you'll have it all in your mind and can then start solving those exciting puzzles. You probably already know the 5's and the 1's and 2's. Go ahead and learn the rest. You can do it!

1's ____ Days	2's ____ Days	3's ____ Days	4's ____ Days	5's ____ Days
1 X 0 = 0	2 X 0 = 0	3 X 0 = 0	4 X 0 = 0	5 X 0 = 0
1 X 1 = 1	2 X 1 = 2	3 X 1 = 3	4 X 1 = 4	5 X 1 = 5
1 X 2 = 2	2 X 2 = 4	3 X 2 = 6	4 X 2 = 8	5 X 2 = 10
1 X 3 = 3	2 X 3 = 6	3 X 3 = 9	4 X 3 = 12	5 X 3 = 15
1 X 4 = 4	2 X 4 = 8	3 X 4 = 12	4 X 4 = 16	5 X 4 = 20
1 X 5 = 5	2 X 5 = 10	3 X 5 = 15	4 X 5 = 20	5 X 5 = 25
1 X 6 = 6	2 X 6 = 12	3 X 6 = 18	4 X 6 = 24	5 X 6 = 30
1 X 7 = 7	2 X 7 = 14	3 X 7 = 21	4 X 7 = 28	5 X 7 = 35
1 X 8 = 8	2 X 8 = 16	3 X 8 = 24	4 X 8 = 32	5 X 8 = 40
1 X 9 = 9	2 X 9 = 18	3 X 9 = 27	4 X 9 = 36	5 X 9 = 45

Remember, when you finish mastering through the 3's, stop and take your "0 – 3" timed test.

The next columns will appear on the next page.

6's ____ Days	7's ____ Days	8's ____ Days	9's ____ Days
6 X 0 = 0	7 X 0 = 0	8 X 0 = 0	9 X 0 = 0
6 X 1 = 6	7 X 1 = 7	8 X 1 = 8	9 X 1 = 9
6 X 2 = 12	7 X 2 = 14	8 X 2 = 16	9 X 2 = 18
6 X 3 = 18	7 X 3 = 21	8 X 3 = 24	9 X 3 = 27
6 X 4 = 24	7 X 4 = 28	8 X 4 = 32	9 X 4 = 36
6 X 5 = 30	7 X 5 = 35	8 X 5 = 40	9 X 5 = 45
6 X 6 = 36	7 X 6 = 42	8 X 6 = 48	9 X 6 = 54
6 X 7 = 42	7 X 7 = 49	8 X 7 = 56	9 X 7 = 63
6 X 8 = 48	7 X 8 = 56	8 X 8 = 64	9 X 8 = 72
6 X 9 = 54	7 X 9 = 63	8 X 9 = 72	9 X 9 = 81

You've finished mastering the multiplication tables, you've taken all four timed tests, and you are now ready to give your mind the exercise it needs. What will result will be lots of fun and enhanced self-esteem, because you are going to succeed with each puzzle. Enjoy them and share them with your friends for years to come.

G. Solution Section

It is possible that you may discover a valid solution that I failed to list, simply because I didn't come up with it. If that happens, and I hope it doesn't happen often, I would like to congratulate you in advance for being that sharp.

Regarding the solutions that follow, I would like to take two examples to illustrate how the solutions are shown.

Example 1: $(9 \times \square) - 1 = \square \times 4$ There is only one solution to this example and it will be shown as 12. No, that is not twelve. Remember, that every empty box will have only a one-digit number as its solution (1 – 9). Therefore, the 12 actually means that "1" is the solution to the first box and "2" is the solution to the second box.

Example 2: $(8 \times \square) - (\square \times 7) - 1 = \square \times 9$ This puzzle has four solutions and they will be shown as 321; 532; 743; 954. With the 954, the 9 represents the first box, the 5, the second box, and the 4, the third box.

Now that you understand how to read the solutions, here they are.

1) 12	13) 78	25) 25	37) 26	49) 13
2) 12	14) 25	26) 56	38) 28	50) 34
3) 15	15) 24	27) 67	39) 35	51) 45
4) 37	16) 28	28) 26	40) 23	52) 54
5) 45	17) 12	29) 76	41) 46	53) 65
6) 35	18) 13	30) 87	42) 56	54) 21
7) 68	19) 12	31) 23	43) 24	55) 32
8) 34	20) 22	32) 25	44) 54	56) 43
9) 47	21) 23	33) 34	45) 65	57) 32
10) 67	22) 22	34) 24	46) 22	58) 42
11) 24	23) 35	35) 43	47) 76	59) 53
12) 22	24) 46	36) 45	48) 86	60) 64

61) 43	84) 218	107) 216	130) 22; 87
62) 54	85) 317	108) 218	131) 22; 67
63) 74	86) 216	109) 128	132) 34; 79
64) 53	87) 129	110) 327	133) 23; 68
65) 64	88) 128	111) 218	134) 24; 49
66) 23, 58	89) 318	112) 219	135) 13; 38
67) 42	90) 217	113) 139	136) 11; 46
68) 53	91) 218	114) 216; 419	137) 24; 59
69) 73	92) 219	115) 116; 319	138) 23; 57
70) 42	93) 219	116) 112; 236; 426; 616	139) 32; 86
71) 32	94) 218	117) 316; 328	140) 22; 56
72) 52	95) 219	118) 13; 27	141) 43; 97
73) 42	96) 219	119) 15; 29	142) 12; 46
74) 31	97) 217	120) 23; 57	143) 34; 68
75) 52	98) 227	121) 34; 68	144) 22; 76
76) 62	99) 139	122) 12; 49	145) 21; 74
77) 52	100) 228	123) 22; 89	146) 32; 85
78) 62	101) 129	124) 12; 59	147) 22; 65
79) 41	102) 128	125) 22; 98	148) 42; 95
80) 62	103) 419	126) 22; 78	149) 32; 75
81) 82	104) 217	127) 32; 75	150) 24; 47
82) 217	105) 217	128) 43; 86	151) 21; 64
83) 129; 217	106) 218	129) 12; 47	152) 21; 94

153) 51; 92

154) 31; 83

155) 31; 72

156) 42; 94

157) 41; 72

158) 21; 73

159) 127; 317

160) 126; 228

161) 127; 229

162) 216; 229

163) 126; 316

164) 417; 429

165) 317; 329

166) 126; 317

167) 316; 328

168) 127; 318

169) 229; 318

170) 216; 228

171) 129; 218

172) 139; 216

173) 127; 216

174) 216; 229

175) 126; 215

176) 137; 417

177) 125; 328

178) 214; 327

179) 126; 329

180) 126; 228

181) 127; 229

182) 149; 227

183) 127; 229

184) 226; 417

185) 126; 419

186) 317; 329

187) 127; 319

188) 315; 338

189) 138; 317

190) 125; 316

191) 139; 318

192) 216; 228

193) 149; 215

194) 138; 317

195) 116; 129

196) 125; 418

197) 113; 328

198) 239; 317

199) 226; 249

200) 126; 228

201) 149; 227

202) 127; 229

203) 137; 215

204) 116; 129

205) 23; 46; 69

206) 12; 25; 38

207) 23; 46; 69

208) 21; 53; 85

209) 12; 35; 58

210) 23; 46; 69

211) 22; 45; 68

212) 21; 53; 85

213) 21; 52; 83

214) 22; 54; 86

215) 216; 228; 419

216) 127; 216; 419

217) 215; 228; 418

218) 128; 216; 419

219) 114; 127; 418

220) 138; 227; 316

221) 126; 215; 519

222) 215; 228; 519

223) 114; 126; 138; 418

224) 136; 326; 516

225) 226; 249; 417

226) 137; 316; 429

227) 125; 238; 417

228) 113; 429; 518

229) 239; 328; 417

230) 138; 215; 329

231) 139; 228; 317

232) 138; 227; 316

233) 215; 227; 239

234) 215; 227; 239

235) 114; 127; 419

236) 224; 347; 627

237) 148; 226; 518

238) 247; 314; 527

239) 125; 238; 518

240) 148; 226; 429

241) 137; 215; 418

242) 136; 214; 428

243) 124; 338; 416

244) 148; 315; 428

245) 138; 215; 418

246) 114; 216; 318

247) 115; 217; 319

248) 236; 427; 618

249) 125; 238; 519

250) 214; 237; 529

251) 225; 248; 517

252) 137; 328; 519

253) 214; 248; 439

254) 148; 316; 429

255) 136; 249; 417

256) 113; 238; 429

257) 159; 214; 327

258) 215; 227; 239

259) 236; 427; 618

260) 214; 237; 518

261) 124; 147; 428

262) 169; 225; 416

263) 135; 326; 517

264) 124; 259; 315

265) 114; 328; 416; 718

266) 415; 427; 439; 818

267) 237; 314; 527; 817

268) 315; 328; 617; 919

269) 226; 427; 628; 829

270) 215; 228; 517; 819

271) 12; 35; 58

272) 13; 25; 37; 49

273) 321; 532; 743; 954

274) 211; 422; 633; 844; 919

275) 215; 227; 239; 418

276) 126; 215; 329; 418

277) 137; 314; 427; 717

278) 113; 439; 528; 617

279) 325; 348; 627; 929

280) 125; 238; 316; 429

281) 137; 215; 328; 519

282) 214; 226; 238; 518

283) 215; 227; 239; 519

284) 113; 239; 328; 417

285) 139; 215; 329; 519

286) 249; 338; 427; 516

287) 314; 337; 628; 919

288) 214; 237; 428; 619

289) 236; 415; 449; 628

290) 214; 226; 238; 619

291) 225; 248; 516; 539

292) 225; 248; 416; 439

293) 136; 249; 315; 428

294) 148; 214; 327; 619

295) 157; 224; 437; 717

296) 135; 348; 415; 628

297) 147; 214; 338; 529

298) 135; 259; 326; 517

299) 168; 314; 426; 538

300) 159; 237; 315; 529

301) 124; 248; 315; 439

302) 158; 225; 349; 416

303) 215; 227; 239; 418

304) 113; 249; 327; 619

305) 113; 237; 428; 619

306) 135; 169; 326; 517

307) 213; 325; 437; 549

308) 179; 314; 426; 538

309) 124; 147; 416; 439

310) 113; 136; 159; 428

311) 124; 158; 315; 349

312) 146; 337; 528; 719

313) 113; 136; 159; 417

314) 125; 326; 527; 728; 929

315) 113; 238; 314; 439; 515; 716; 917

316) 149; 225; 426; 627; 828

317) 214; 227; 516; 529; 818

318) 126; 214; 428; 516; 818

319) 423; 616; 653; 846; 883

320) 432; 635; 661; 838; 864

321) 514; 634; 754; 874; 994

322) 211; 422; 633; 844; 991

323) 312; 523; 734; 881; 945

324) 541; 643; 745; 847; 949

325) 125; 226; 327; 428; 529

326) 124; 337; 414; 627; 917

327) 235; 269; 436; 637; 838

328) 125; 238; 415; 528; 818

329) 314; 326; 338; 717; 729

330) 157; 324; 447; 614; 737

331) 148; 225; 438; 515; 728

332) 214; 226; 238; 617; 629

333) 113; 138; 227; 316; 519

334) 214; 226; 238; 417; 429

335) 145; 346; 547; 748; 949

336) 136; 214; 249; 327; 518

337) 112; 246; 436; 626; 816

338) 147; 213; 337; 527; 717

339) 224; 247; 526; 549; 828

340) 148; 237; 326; 415; 919

341) 136; 225; 314; 729; 818

342) 124; 213; 539; 628; 717

343) 124; 147; 426; 449; 728

344) 213; 325; 437; 549; 918

345) 224; 258; 437; 616; 829

346) 213; 247; 426; 639; 818

347) 213; 236; 259; 527; 818

348) 146; 258; 515; 627; 739

349) 124; 147; 415; 438; 729

350) 157; 269; 414; 526; 638

351) 124; 147; 315; 338; 529

352) 113; 125; 137; 149; 418

353) 166; 234; 468; 536; 838

354) 112; 224; 336; 448; 818

355) 145; 268; 425; 548; 828

356) 134; 257; 414; 537; 817

357) 134; 246; 358; 616; 728

358) 112; 224; 336; 448; 919

359) 134; 246; 358; 717; 829

360) 123; 235; 347; 459; 818

361) 124; 236; 348; 514; 626; 738

362) 113; 239; 327; 415; 629; 717

363) 113; 126; 139; 415; 428; 717

364) 113; 226; 339; 415; 528; 717

365) 314; 326; 338; 616; 628; 918

366) 126; 239; 315; 428; 617; 919

367) 138; 226; 314; 528; 616; 918

368) 139; 227; 315; 529; 617; 919

369) 324; 347; 625; 648; 926; 949

370) 214; 226; 238; 516; 528; 818

371) 148; 225; 349; 426; 627; 828

372) 138; 214; 327; 516; 629; 818

373) 224; 247; 525; 548; 826; 849

374) 148; 213; 337; 526; 715; 839

375) 136; 325; 449; 514; 638; 827

376) 124; 147; 314; 337; 527; 717

377) 113; 249; 338; 427; 516; 919

378) 159; 248; 337; 426; 515; 929

379) 112; 259; 348; 437; 526; 615

380) 134; 357; 424; 647; 714; 937

381) 113; 226; 339; 516; 629; 919

382) 224; 247; 414; 437; 627; 817

383) 225; 248; 415; 438; 628; 818

384) 148; 237; 326; 415; 729; 818

385) 224; 258; 425; 459; 626; 827

386) 134; 257; 424; 547; 714; 837

387) 167; 234; 457; 524; 747; 814

388) 136; 225; 314; 539; 628; 717

389) 124; 213; 449; 538; 627; 716

390) 124; 213; 349; 438; 527; 616

391) 213; 225; 237; 249; 616; 628

392) 112; 235; 358; 525; 648; 815; 938

393) 112; 248; 325; 538; 615; 828

394) 113; 125; 137; 149; 417; 429

395) 123; 257; 313; 447; 637; 827

396) 123; 358; 436; 514; 749; 827

397) 124; 159; 237; 315; 428; 619

398) 146; 258; 314; 426; 538; 818

399) 123; 246; 369; 425; 548; 727

400) 158; 236; 314; 449; 527; 818

401) 146; 224; 359; 437; 515; 728

402) 168; 224; 347; 526; 649; 828

403) 135; 213; 348; 426; 639; 717

404) 122; 266; 312; 456; 646; 836

405) 124; 248; 314; 438; 628; 818

406) 156; 212; 346; 536; 726; 916

407) 166; 212; 356; 546; 736; 926

408) 134; 279; 335; 536; 737; 938

409) 112; 269; 347; 425; 738; 816

410) 213; 236; 259; 515; 538; 817

411) 157; 235; 313; 548; 626; 939

412) 146; 224; 459; 537; 615; 928

413) 157; 269; 314; 426; 538; 919

414) 179; 224; 347; 515; 638; 929

415) 135; 169; 314; 348; 527; 919

416) 134; 268; 313; 447; 626; 939

417) 112; 224; 336; 448; 817; 929

418) 123; 257; 436; 615; 749; 928

419) 112; 246; 425; 559; 738; 917

420) 168; 213; 325; 437; 549; 818

421) 134; 257; 425; 548; 716; 839

422) 123; 246; 369; 414; 537; 828

423) 112; 235; 358; 526; 649; 817

424) 156; 335; 469; 514; 648; 827

425) 123; 235; 347; 459; 716; 828

426) 145; 279; 324; 458; 637; 816

427) 134; 246; 358; 515; 627; 739

428) 123; 235; 347; 459; 616; 728

429) 145; 268; 313; 436; 559; 727

430) 112; 146; 325; 359; 538; 717

431) 112; 135; 158; 426; 449; 717

432) 188; 256; 324; 558; 626; 928

433) 179; 213; 325; 437; 549; 919

434) 123; 246; 369; 526; 649; 929

435) 146; 213; 359; 426; 639; 919

436) 112; 235; 358; 515; 638; 918

437) 188; 267; 346; 425; 749; 828

438) 123; 235; 347; 459; 717; 829

439) 133; 266; 399; 435; 568; 737

440) 166; 245; 324; 569; 648; 727

441) 144; 223; 389; 468; 547; 626

442) 199; 278; 357; 436; 515; 929

443) 188; 267; 346; 425; 839; 918

444) 177; 256; 335; 414; 749; 828

445) 166; 245; 324; 659; 738; 817

446) 155; 234; 313; 569; 648; 727

447) 322; 543; 616; 764; 837; 985

448) 313; 543; 626; 773; 856; 939

449) 331; 534; 662; 737; 865; 993

450) 542; 634; 726; 818; 891; 983

451) 441; 533; 625; 717; 882; 974

452) 432; 524; 616; 781; 873; 965

453) 331; 423; 515; 772; 864; 956

454) 551; 643; 735; 827; 919; 992

455) 313; 432; 551; 727; 846; 965

456) 423; 542; 661; 718; 837; 956

457) 414; 533; 652; 771; 828; 947

458) 524; 643; 762; 819; 881; 938

459) 515; 634; 753; 872; 929; 991

460) 331; 413; 662; 744; 826; 993

461) 312; 561; 643; 725; 892; 974

462) 423; 542; 661; 727; 846; 965

463) 321; 625; 642; 929; 946; 963

464) 313; 431; 626; 744; 862; 939

465) 423; 452; 481; 716; 745; 774

466) 713; 724; 735; 746; 757; 768; 779

467) 613; 624; 635; 646; 657; 668; 679

468) 125; 148; 226; 249; 327; 428; 529

469) 513; 524; 535; 546; 557; 568; 579

470) 124; 248; 325; 449; 526; 727; 928

471) 125; 238; 314; 427; 616; 729; 918

472) 167; 245; 323; 568; 646; 724; 969

473) 168; 257; 346; 435; 524; 613; 959

474) 157; 246; 335; 424; 513; 859; 948

475) 145; 223; 468; 546; 624; 869; 947

476) 124; 248; 313; 437; 626; 815; 939

477) 159; 224; 348; 413; 537; 726; 915

478) 147; 224; 348; 425; 549; 626; 827

479) 137; 213; 326; 439; 515; 628; 817

480) 189; 267; 345; 423; 668; 746; 824

481) 137; 213; 338; 414; 539; 615; 816

482) 813; 824; 835; 846; 857; 868; 879

483) 112; 224; 336; 448; 614; 726; 838

484) 179; 268; 357; 446; 535; 624; 713

485) 123; 146; 169; 424; 447; 725; 748

486) 112; 236; 425; 549; 614; 738; 927

487) 113; 148; 237; 326; 415; 539; 717

488) 157; 224; 347; 414; 537; 727; 917

489) 134; 268; 335; 469; 536; 737; 938

490) 135; 247; 359; 414; 526; 638; 917

491) 147; 236; 325; 414; 739; 828; 917

492) 113; 125; 137; 149; 516; 528; 919

493) 145; 212; 368; 435; 658; 725; 948

494) 136; 213; 349; 426; 639; 716; 929

495) 134; 257; 324; 447; 514; 637; 827

496) 158; 213; 325; 437; 549; 716; 828

497) 133; 266; 399; 434; 567; 735; 868

498) 134; 279; 346; 413; 558; 625; 837

499) 123; 268; 335; 547; 614; 759; 826

500) 135; 224; 313; 549; 638; 727; 816

501) 178; 245; 312; 468; 535; 758; 825

502) 123; 235; 347; 459; 514; 626; 738

503) 113; 125; 137; 149; 316; 328; 519

504) 112; 169; 247; 325; 538; 616; 829

505) 112; 135; 158; 414; 437; 716; 739

506) 134; 246; 358; 414; 526; 638; 918

507) 144; 288; 345; 489; 546; 747; 948

508) 145; 268; 324; 447; 626; 749; 928

509) 112; 169; 247; 325; 538; 616; 829

510) 133; 277; 323; 467; 513; 657; 847

511) 134; 268; 324; 458; 514; 648; 838

512) 159; 225; 349; 415; 539; 729; 919

513) 123; 268; 324; 469; 525; 726; 927

514) 168; 246; 324; 559; 637; 715; 828

515) 167; 223; 368; 424; 569; 625; 826

516) 133; 212; 388; 467; 546; 625; 959

517) 177; 234; 389; 446; 658; 715; 927

518) 166; 223; 378; 435; 647; 859; 916

519) 223; 266; 435; 478; 647; 816; 859

520) 177; 245; 313; 468; 536; 759; 827

521) 166; 234; 389; 457; 525; 748; 816

522) 155; 212; 367; 424; 579; 636; 848

523) 144; 299; 356; 413; 568; 625; 837

524) 133; 288; 345; 557; 614; 769; 826

525) 177; 256; 335; 414; 569; 648; 727

526) 166; 245; 324; 479; 558; 637; 716

527) 155; 223; 378; 446; 514; 669; 737

528) 112; 179; 224; 336; 448; 717; 829

529) 122; 255; 388; 424; 557; 726; 859

530) 187; 223; 356; 489; 525; 658; 827

531) 177; 256; 335; 414; 659; 738; 817

532) 155; 223; 389; 457; 525; 759; 827

533) 144; 212; 378; 446; 514; 748; 816

534) 155; 234; 313; 479; 558; 637; 716

535) 113; 125; 137; 149; 316; 328; 519

536) 462; 515; 592; 645; 775; 828; 958

537) 313; 443; 573; 626; 756; 886; 939

538) 332; 471; 544; 683; 756; 895; 968

539) 221; 313; 562; 654; 746; 838; 995

540) 323; 452; 526; 655; 784; 858; 987

541) 313; 461; 534; 682; 755; 828; 976

542) 221; 442; 515; 663; 736; 884; 957

543) 423; 571; 644; 717; 792; 865; 938

544) 341; 414; 562; 635; 783; 856; 929

545) 322; 515; 552; 745; 782; 938; 975

546) 322; 552; 635; 718; 782; 865; 948

547) 322; 525; 653; 728; 781; 856; 984

548) 212; 331; 526; 645; 764; 883; 959

549) 221; 424; 552; 627; 755; 883; 958

550) 331; 434; 537; 662; 765; 868; 993

551) 322; 525; 551; 728; 754; 957; 983

552) 432; 517; 644; 729; 771; 856; 983

553) 212; 424; 551; 636; 763; 848; 975

554) 322; 534; 619; 661; 746; 873; 958

555) 391; 492; 593; 694; 795; 896; 997

556) 212; 341; 415; 544; 673; 747; 876

557) 231; 443; 516; 582; 655; 728; 794

558) 313; 332; 351; 727; 746; 765; 784

559) 314; 441; 526; 653; 738; 865; 992

560) 332; 472; 524; 664; 716; 856; 996

561) 414; 534; 654; 774; 828; 894; 948

562) 322; 451; 515; 644; 773; 837; 966

563) 423; 552; 616; 681; 745; 874; 938

564) 441; 543; 645; 747; 849; 882; 984

565) 313; 432; 551; 617; 736; 855; 974

566) 322; 415; 551; 644; 737; 873; 966

567) 12; 23; 34; 45; 56; 67; 78; 89

568) 112; 135; 158; 413; 436; 459; 714; 737

569) 114; 128; 215; 229; 316; 417; 518; 619

570) 512; 523; 534; 545; 556; 567; 578; 589

571) 412; 423; 434; 445; 456; 467; 478; 489

572) 136; 213; 337; 414; 538; 615; 739; 816

573) 412; 423; 434; 445; 456; 467; 478; 489

574) 121; 242; 363; 484; 643; 764; 885; 923

575) 187; 222; 344; 466; 588; 745; 867; 989

576) 123; 268; 346; 424; 569; 647; 725; 948

577) 178; 256; 334; 412; 568; 646; 724; 958

578) 158; 212; 324; 436; 548; 714; 826; 938

579) 156; 234; 312; 468; 546; 624; 858; 936

580) 147; 259; 313; 425; 537; 649; 815; 927

581) 144; 266; 388; 412; 534; 656; 778; 924

582) 135; 247; 359; 413; 525; 637; 749; 915

583) 134; 212; 368; 446; 524; 758; 836; 914

584) 156; 234; 312; 479; 557; 635; 713; 958

585) 134; 223; 312; 569; 658; 747; 836; 925

586) 123; 212; 469; 558; 647; 736; 825; 914

587) 145; 223; 368; 446; 524; 669; 747; 825

588) 135; 224; 313; 459; 548; 637; 726; 815

589) 122; 244; 366; 488; 512; 634; 756; 878

590) 212; 235; 258; 513; 536; 559; 814; 837

591) 178; 256; 334; 412; 579; 657; 735; 813

592) 146; 258; 313; 425; 537; 649; 816; 928

593) 123; 257; 324; 458; 525; 659; 726; 927

594) 112; 159; 248; 337; 426; 515; 829; 918

595) 167; 234; 379; 446; 513; 658; 725; 937

596) 156; 223; 368; 435; 647; 714; 859; 926

597) 123; 157; 324; 358; 525; 559; 726; 927

598) 155; 288; 323; 456; 589; 624; 757; 925

599) 145; 212; 357; 424; 569; 636; 848; 915

600) 156; 223; 379; 446; 513; 669; 736; 959

601) 123; 246; 369; 413; 536; 659; 826; 949

602) 168; 212; 335; 458; 625; 715; 748; 915

603) 135; 247; 314; 359; 426; 538; 717; 829

604) 244; 287; 434; 477; 624; 667; 814; 857

605) 144; 277; 334; 467; 524; 657; 714; 847

606) 213; 236; 259; 426; 449; 616; 639; 829

607) 165; 233; 398; 466; 534; 699; 767; 835

608) 156; 223; 357; 424; 558; 625; 759; 826

609) 122; 255; 388; 423; 556; 689; 724; 857

610) 212; 246; 413; 447; 614; 648; 815; 849

611) 146; 213; 258; 325; 437; 549; 616; 728

612) 312; 344; 376; 624; 656; 688; 936; 968

613) 157; 213; 269; 325; 437; 549; 717; 829

614) 123; 179; 235; 347; 459; 515; 627; 739

615) 166; 234; 378; 446; 514; 658; 726; 938

616) 112; 179; 235; 358; 414; 537; 716; 839

617) 165; 287; 312; 434; 556; 678; 825; 947

618) 133; 277; 334; 478; 535; 679; 736; 937

619) 134; 257; 313; 436; 559; 615; 738; 917

620) 133; 255; 377; 499; 524; 646; 768; 915

621) 122; 244; 366; 488; 513; 635; 757; 879

622) 166; 223; 367; 424; 568; 625; 769; 826

623) 144; 288; 334; 478; 524; 668; 714; 858

624) 145; 279; 335; 469; 525; 659; 715; 849

625) 156; 212; 357; 413; 558; 614; 759; 815

626) 323; 355; 387; 614; 646; 678; 937; 969

627) 155; 234; 313; 389; 468; 547; 626; 939

628) 122; 298; 377; 456; 535; 614; 869; 948

629) 133; 288; 356; 424; 579; 647; 715; 938

630) 187; 266; 345; 424; 679; 758; 837; 916

631) 212; 244; 276; 535; 567; 599; 826; 858

632) 212; 255; 298; 424; 467; 636; 679; 848

633) 244; 287; 413; 456; 499; 625; 668; 837

634) 176; 255; 334; 413; 589; 668; 747; 826

635) 165; 244; 323; 499; 578; 657; 736; 815

636) 122; 154; 186; 413; 445; 477; 736; 768

637) 176; 212; 345; 478; 514; 647; 816; 949

638) 144; 277; 313; 446; 579; 615; 748; 917

639) 177; 245; 313; 479; 547; 615; 849; 917

640) 323; 453; 516; 583; 646; 776; 839; 969

641) 212; 342; 472; 525; 655; 785; 838; 968

642) 212; 351; 424; 563; 636; 775; 848; 987

643) 332; 461; 535; 664; 738; 793; 867; 996

644) 212; 461; 553; 645; 737; 829; 894; 986

645) 442; 534; 626; 691; 718; 783; 875; 967

646) 341; 433; 525; 617; 682; 774; 866; 958

647) 332; 424; 516; 581; 673; 765; 857; 949

648) 221; 451; 534; 617; 681; 764; 847; 994

649) 331; 524; 561; 717; 754; 791; 947; 984

650) 323; 416; 553; 646; 739; 783; 876; 969

651) 331; 414; 561; 644; 727; 791; 874; 957

652) 314; 433; 552; 628; 671; 747; 866; 985

653) 313; 441; 516; 644; 719; 772; 847; 975

654) 221; 416; 535; 654; 773; 849; 892; 968

655) 212; 415; 543; 618; 671; 746; 874; 949

656) 313; 516; 542; 719; 745; 771; 948; 974

657) 212; 251; 424; 463; 636; 675; 848; 887

658) 231; 361; 414; 544; 674; 727; 857; 987

659) 231; 434; 462; 637; 665; 693; 868; 896

660) 414; 433; 452; 471; 828; 847; 866; 885

661) 231; 334; 437; 462; 565; 668; 693; 899

662) 213; 426; 443; 639; 656; 673; 869; 886

663) 332; 435; 538; 561; 664; 767; 893; 996

664) 417; 425; 433; 441; 858; 866; 874; 882

665) 221; 314; 451; 544; 637; 681; 774; 867

666) 221; 315; 442; 536; 663; 757; 884; 978

667) 241; 381; 433; 573; 625; 765; 817; 957

668) 322; 351; 615; 644; 673; 937; 966; 995

669) 221; 341; 461; 581; 615; 735; 855; 975

670) 212; 441; 534; 627; 763; 856; 949; 992

671) 315; 322; 637; 644; 651; 959; 966; 973

672) 332; 481; 563; 645; 727; 794; 876; 958

673) 311; 342; 373; 622; 653; 684; 933; 964; 995

674) 183; 221; 342; 463; 584; 622; 743; 864; 985

675) 142; 263; 384; 422; 543; 664; 785; 823; 944

676) 131; 162; 193; 411; 442; 473; 722; 753; 784

677) 121; 152; 183; 432; 463; 494; 712; 743; 774

678) 142; 284; 332; 474; 522; 664; 712; 854; 996

679) 121; 263; 311; 453; 595; 643; 785; 833; 975

680) 194; 242; 384; 432; 574; 622; 764; 812; 954

681) 184; 232; 374; 422; 564; 612; 754; 896; 944

682) 163; 211; 353; 495; 543; 685; 733; 875; 923

683) 112; 148; 225; 338; 415; 528; 617; 718; 819

684) 153; 211; 364; 422; 575; 633; 786; 844; 997

685) 185; 264; 343; 422; 586; 665; 744; 823; 987

686) 132; 285; 343; 496; 554; 612; 765; 823; 976

687) 164; 243; 322; 486; 565; 644; 723; 887; 966

688) 195; 253; 311; 464; 522; 675; 733; 886; 944

689) 143; 222; 386; 465; 544; 623; 787; 866; 945

690) 174; 232; 385; 443; 596; 654; 712; 865; 923

691) 136; 225; 314; 349; 438; 527; 616; 829; 918

692) 322; 354; 386; 612; 644; 676; 934; 966; 998

693) 112; 124; 136; 148; 414; 426; 438; 716; 728

694) 132; 164; 196; 422; 454; 486; 712; 744; 776

695) 113; 125; 137; 149; 415; 427; 439; 717; 729

696) 197; 211; 322; 433; 544; 655; 766; 877; 988

697) 165; 276; 387; 498; 512; 623; 734; 845; 956

698) 123; 212; 269; 358; 447; 536; 625; 714; 949

699) 144; 266; 388; 423; 545; 667; 789; 824; 946

700) 167; 245; 323; 468; 546; 624; 769; 847; 925

701) 157; 246; 335; 424; 513; 659; 748; 837; 926

702) 122; 244; 366; 488; 523; 645; 767; 889; 924

703) 146; 235; 324; 413; 559; 648; 737; 826; 915

704) 122; 233; 344; 455; 566; 677; 788; 899; 913

705) 112; 169; 258; 347; 436; 525; 614; 849; 938

706) 111; 233; 355; 477; 599; 623; 745; 867; 989

707) 198; 222; 344; 466; 588; 612; 734; 856; 978

708) 167; 245; 323; 479; 557; 635; 713; 869; 947

709) 145; 223; 379; 457; 535; 613; 769; 847; 925

710) 133; 255; 377; 499; 523; 645; 767; 889; 913

711) 211; 242; 273; 522; 553; 584; 833; 864; 895

712) 221; 252; 283; 532; 563; 594; 812; 843; 874

713) 146; 224; 269; 347; 425; 548; 626; 749; 827

714) 222; 254; 286; 512; 544; 576; 834; 866; 898

715) 211; 243; 275; 533; 565; 597; 823; 855; 887

716) 213; 225; 237; 249; 515; 527; 539; 817; 829

717) 112; 168; 246; 324; 458; 536; 614; 748; 826

718) 414; 425; 436; 447; 458; 469; 817; 828; 839

719) 163; 232; 395; 464; 533; 696; 765; 834; 997

720) 142; 211; 374; 443; 512; 675; 744; 813; 976

721) 121; 284; 353; 422; 585; 654; 723; 886; 955

722) 184; 253; 322; 485; 554; 623; 786; 855; 924

723) 132; 264; 396; 433; 565; 697; 734; 866; 998

724) 185; 222; 354; 486; 523; 655; 787; 824; 956

725) 174; 211; 343; 475; 512; 644; 776; 813; 945

726) 143; 254; 365; 476; 587; 698; 724; 835; 946

727) 153; 285; 322; 454; 586; 623; 755; 887; 924

728) 112; 168; 235; 358; 425; 548; 615; 738; 928

729) 112; 135; 158; 325; 348; 515; 538; 728; 918

730) 154; 222; 387; 455; 523; 688; 756; 824; 989

731) 122; 287; 355; 423; 589; 656; 724; 889; 957

732) 144; 277; 323; 456; 589; 635; 768; 814; 947

733) 133; 266; 312; 399; 445; 578; 624; 757; 936

734) 122; 255; 388; 434; 567; 613; 746; 879; 925

735) 176; 244; 312; 477; 545; 613; 778; 846; 914

736) 144; 277; 312; 445; 578; 613; 746; 879; 914

737) 112; 146; 313; 347; 514; 548; 715; 749; 916

738) 412; 433; 454; 475; 496; 824; 845; 866; 887

739) 233; 276; 423; 466; 613; 656; 699; 846; 889

740) 222; 265; 412; 455; 498; 645; 688; 835; 878

741) 145; 212; 268; 335; 458; 525; 648; 715; 838

742) 155; 288; 334; 467; 513; 646; 779; 825; 958

743) 322; 343; 364; 385; 713; 734; 755; 776; 797

744) 222; 243; 264; 285; 613; 634; 655; 676; 697

745) 211; 232; 253; 274; 295; 623; 644; 665; 686

746) 185; 296; 312; 423; 534; 645; 756; 867; 978

747) 143; 286; 344; 487; 545; 688; 746; 889; 947

748) 144; 223; 288; 367; 446; 525; 669; 748; 827

749) 164; 275; 386; 497; 513; 624; 735; 846; 957

750) 153; 264; 375; 486; 597; 613; 724; 835; 946

751) 122; 265; 323; 466; 524; 687; 725; 868; 926

752) 132; 243; 354; 465; 576; 687; 798; 814; 925

753) 166; 245; 324; 389; 468; 547; 626; 849; 928

754) 175; 233; 376; 434; 577; 635; 778; 836; 979

755) 164; 222; 365; 423; 566; 624; 767; 825; 968

756) 233; 265; 297; 513; 545; 577; 825; 857; 889

757) 222; 254; 286; 534; 566; 598; 814; 846; 878

758) 144; 212; 288; 356; 424; 568; 636; 848; 916

759) 133; 277; 345; 413; 489; 557; 625; 769; 837

760) 122; 198; 266; 334; 478; 546; 614; 758; 826

761) 176; 298; 323; 445; 567; 689; 714; 836; 958

762) 122; 266; 323; 467; 524; 668; 725; 869; 926

763) 155; 299; 345; 489; 535; 679; 725; 869; 915

764) 122; 154; 186; 434; 466; 498; 714; 746; 778

765) 133; 176; 345; 388; 514; 557; 726; 769; 938

766) 122; 165; 334; 377; 546; 589; 715; 758; 927

767) 144; 223; 299; 378; 457; 536; 615; 849; 928

768) 133; 212; 288; 367; 446; 525; 759; 838; 917

769) 144; 212; 299; 367; 435; 536; 658; 726; 949

770) 233; 265; 297; 524; 556; 588; 815; 847; 879

771) 111; 143; 175; 434; 466; 498; 725; 757; 789

772) 313; 352; 391; 525; 564; 737; 776; 949; 988

773) 222; 352; 415; 482; 545; 675; 738; 868; 998

774) 231; 361; 414; 491; 544; 674; 727; 857; 987

775) 323; 351; 526; 554; 582; 729; 757; 785; 988

776) 515; 534; 553; 572; 591; 929; 948; 967; 986

777) 519; 527; 535; 543; 551; 968; 976; 984; 992

778) 213; 316; 419; 441; 544; 647; 772; 875; 978

779) 261; 372; 415; 483; 526; 594; 637; 748; 859

780) 121; 232; 343; 454; 565; 676; 719; 787; 898

781) 213; 343; 426; 473; 556; 639; 686; 769; 899

782) 231; 361; 424; 491; 554; 617; 684; 747; 877

783) 222; 425; 453; 481; 628; 656; 684; 859; 887

784) 213; 324; 435; 481; 546; 592; 657; 768; 879

785) 216; 223; 439; 446; 453; 669; 676; 683; 899

786) 323; 417; 544; 638; 671; 765; 859; 892; 986

787) 121; 242; 315; 363; 436; 484; 557; 678; 799

788) 121; 324; 353; 382; 527; 556; 585; 759; 788

789) 223; 316; 353; 446; 483; 539; 576; 669; 799

790) 232; 335; 361; 438; 464; 567; 593; 696; 799

791) 332; 425; 518; 562; 655; 748; 792; 885; 978

792) 232; 261; 435; 464; 493; 638; 667; 696; 899

793) 222; 261; 434; 473; 646; 685; 819; 858; 897

794) 223; 271; 344; 392; 417; 465; 538; 586; 659

795) 121; 214; 251; 344; 381; 437; 474; 567; 697

796) 112; 215; 241; 318; 344; 447; 473; 576; 679

797) 231; 342; 453; 518; 564; 629; 675; 786; 897

798) 231; 342; 418; 453; 529; 564; 675; 786; 897

799) 131; 234; 262; 337; 365; 393; 468; 496; 599

800) 111; 222; 333; 444; 555; 666; 777; 888; 999

801) 111; 241; 371; 434; 564; 627; 694; 757; 887

802) 232; 335; 361; 438; 464; 567; 593; 696; 799

803) 121; 242; 315; 363; 436; 484; 557; 678; 799

804) 251; 362; 473; 516; 584; 627; 695; 738; 849

805) 213; 261; 334; 382; 455; 528; 576; 649; 697

806) 214; 233; 252; 271; 428; 447; 466; 485; 699

807) 111; 323; 362; 535; 574; 747; 786; 959; 998

808) 222; 362; 434; 574; 646; 718; 786; 858; 998

809) 212; 352; 424; 492; 564; 636; 776; 848; 988

810) 212; 361; 443; 525; 592; 674; 756; 838; 987

811) 323; 342; 361; 627; 646; 665; 684; 969; 988

812) 332; 414; 481; 563; 645; 727; 794; 866; 958

813) 231; 313; 462; 544; 626; 693; 775; 857; 939

814) 314; 434; 554; 628; 674; 748; 794; 868; 988

815) 341; 443; 545; 647; 682; 749; 784; 886; 988

816) 313; 451; 534; 617; 672; 755; 838; 893; 976

817) 212; 433; 516; 571; 654; 737; 792; 875; 958

818) 241; 343; 445; 482; 547; 584; 649; 686; 788

819) 231; 313; 462; 544; 626; 693; 775; 857; 939

820) 232; 334; 371; 436; 473; 538; 575; 677; 779

821) 212; 361; 443; 525; 592; 674; 756; 838; 987

822) 332; 414; 481; 563; 645; 727; 794; 876; 958

823) 124; 158; 236; 314; 348; 426; 538; 616; 728; 918

824) 135; 224; 259; 313; 348; 526; 615; 739; 828; 917

825) 112; 147; 236; 325; 414; 449; 538; 627; 716; 929

826) 144; 266; 312; 388; 434; 556; 678; 724; 846; 968

827) 156; 234; 312; 368; 446; 524; 658; 736; 814; 948

828) 134; 212; 268; 346; 424; 558; 636; 714; 848; 926

829) 112; 159; 236; 313; 437; 514; 638; 715; 839; 916

830) 134; 212; 279; 357; 435; 513; 658; 736; 814; 959

831) 156; 234; 312; 379; 457; 535; 613; 758; 836; 914

832) 313; 335; 357; 379; 615; 637; 659; 917; 928; 939

833) 222; 243; 264; 285; 523; 544; 565; 586; 824; 845; 866; 887

834) 124; 159; 213; 248; 337; 426; 515; 639; 728; 817

835) 175; 286; 312; 397; 423; 534; 645; 756; 867; 978

836) 164; 275; 386; 412; 497; 523; 634; 745; 856; 967

837) 132; 243; 354; 465; 576; 687; 713; 798; 824; 935

838) 123; 168; 235; 347; 414; 459; 526; 638; 817; 929

839) 112; 157; 224; 269; 336; 448; 515; 627; 739; 918

840) 155; 212; 288; 345; 478; 535; 668; 725; 858; 915

841) 145; 212; 279; 346; 413; 547; 614; 748; 815; 949

842) 112; 179; 246; 313; 447; 514; 648; 715; 849; 916

843) 134; 179; 246; 313; 358; 425; 537; 649; 716; 828

844) 165; 212; 287; 334; 456; 578; 625; 747; 869; 916

845) 154; 276; 323; 398; 445; 567; 614; 689; 736; 858

846) 122; 187; 266; 345; 424; 489; 568; 647; 726; 949

847) 133; 255; 377; 424; 499; 546; 668; 715; 837; 959

848) 155; 223; 299; 367; 435; 579; 647; 715; 859; 927

849) 155; 212; 299; 356; 413; 557; 614; 758; 815; 959

850) 241; 352; 416; 463; 527; 574; 638; 685; 749; 796

851) 314; 361; 425; 472; 536; 583; 647; 694; 758; 869

852) 173; 211; 294; 332; 453; 574; 612; 695; 733; 854; 975

853) 152; 273; 311; 394; 432; 553; 674; 712; 795; 833; 954

854) 173; 242; 311; 384; 453; 522; 595; 664; 733; 875; 944

855) 163; 232; 374; 443; 512; 585; 654; 723; 796; 865; 934

856) 132; 264; 322; 396; 454; 512; 586; 644; 776; 834; 966

857) 112; 146; 224; 258; 336; 414; 448; 526; 638; 716; 828

858) 174; 232; 364; 422; 496; 554; 612; 686; 744; 876; 934

859) 154; 265; 376; 412; 487; 523; 598; 634; 745; 856; 967

860) 143; 254; 365; 476; 512; 587; 623; 698; 734; 845; 956

861) 145; 223; 268; 346; 424; 469; 547; 625; 748; 826; 949

862) 132; 211; 296; 375; 454; 533; 612; 697; 776; 855; 934

863) 123; 168; 246; 324; 369; 447; 525; 648; 726; 849; 927

864) 144; 266; 323; 388; 445; 567; 624; 689; 746; 868; 925

865) 133; 255; 377; 423; 499; 545; 667; 713; 789; 835; 957

866) 122; 198; 244; 366; 412; 488; 534; 656; 778; 824; 946

867) 145; 223; 279; 357; 435; 513; 569; 647; 725; 859; 937

868) 176; 222; 298; 344; 466; 512; 588; 634; 756; 878; 924

869) 123; 179; 257; 335; 413; 469; 547; 625; 759; 837; 915

870) 213; 225; 237; 249; 423; 438; 615; 627; 639; 816; 828

871) 164; 275; 312; 386; 423; 497; 534; 645; 756; 867; 978

872) 153; 264; 375; 412; 486; 523; 597; 634; 745; 856; 967

873) 143; 254; 365; 476; 513; 587; 624; 698; 735; 846; 957

874) 132; 243; 354; 465; 576; 613; 687; 724; 798; 835; 946

875) 122; 197; 276; 355; 434; 513; 588; 667; 746; 825; 979

876) 155; 223; 288; 356; 424; 489; 557; 625; 758; 826; 959

877) 165; 244; 323; 398; 477; 556; 635; 714; 789; 868; 947

878) 154; 233; 312; 387; 466; 545; 624; 699; 778; 857; 936

879) 133; 198; 266; 334; 399; 467; 535; 668; 736; 869; 937

880) 143; 222; 297; 376; 455; 534; 613; 688; 767; 846; 925

881) 413; 424; 435; 446; 457; 468; 479; 916; 927; 938; 949

882) 133; 266; 323; 399; 456; 513; 589; 646; 779; 836; 969

883) 122; 165; 312; 355; 398; 545; 588; 735; 778; 925; 968

884) 122; 198; 255; 312; 388; 445; 578; 635; 768; 825; 958

885) 123; 179; 246; 313; 369; 436; 559; 626; 749; 816; 939

886) 233; 265; 297; 434; 466; 498; 635; 667; 699; 836; 868

887) 155; 234; 299; 313; 378; 457; 536; 615; 759; 838; 917

888) 133; 198; 212; 277; 356; 435; 514; 579; 658; 737; 816

889) 122; 197; 244; 366; 413; 488; 535; 657; 779; 826; 948

890) 121; 142; 163; 184; 422; 443; 464; 485; 723; 744; 765; 786

891) 123; 146; 169; 224; 247; 325; 348; 426; 449; 527; 628; 729

892) 113; 136; 159; 214; 237; 315; 338; 416; 439; 517; 618; 719

893) 142; 211; 284; 353; 422; 495; 564; 633; 775; 844; 913; 986

894) 322; 343; 364; 385; 623; 644; 665; 686; 924; 945; 966; 987

895) 121; 194; 263; 332; 474; 543; 612; 685; 754; 823; 896; 965

896) 164; 222; 296; 354; 412; 486; 544; 676; 734; 866; 924; 998

897) 153; 211; 285; 343; 475; 533; 665; 723; 797; 855; 913; 987

898) 112; 134; 156; 178; 324; 346; 368; 514; 536; 558; 726; 748

899) 114; 126; 138; 215; 227; 239; 316; 328; 417; 429; 518; 619

900) 175; 211; 286; 322; 397; 433; 544; 655; 766; 877; 913; 988

901) 122; 187; 244; 366; 423; 488; 545; 667; 724; 789; 846; 968

902) 165; 222; 287; 344; 466; 523; 588; 645; 767; 824; 889; 946

903) 122; 197; 233; 344; 455; 566; 677; 713; 788; 824; 899; 935

904) 112; 157; 235; 313; 358; 436; 514; 559; 637; 715; 838; 916

905) 413; 424; 435; 446; 457; 468; 479; 915; 926; 937; 948; 959

906) 155; 234; 288; 313; 367; 446; 525; 579; 658; 737; 816; 949

907) 144; 212; 277; 345; 413; 478; 546; 614; 679; 747; 815; 948

908) 122; 187; 255; 323; 388; 456; 524; 589; 657; 725; 858; 926

909) 222; 254; 286; 423; 455; 487; 624; 656; 688; 825; 857; 889

910) 212; 244; 276; 413; 445; 477; 614; 646; 678; 815; 847; 879

911) 134; 168; 212; 246; 324; 358; 436; 514; 548; 626; 738; 816; 928

912) 135; 169; 213; 247; 325; 359; 437; 515; 549; 627; 739; 817; 929

913) 134; 179; 212; 257; 335; 413; 458; 536; 614; 659; 737; 815; 938

914) 144; 198; 223; 277; 356; 435; 489; 514; 568; 647; 726; 859; 938

915) 122; 176; 255; 334; 388; 413; 467; 546; 625; 679; 758; 837; 916

916) 135; 169; 213; 247; 325; 359; 437; 515; 549; 627; 739; 817; 929

917) 144; 198; 212; 266; 334; 388; 456; 524; 578; 646; 714; 768; 836; 958

918) 123; 157; 235; 269; 313; 347; 425; 459; 537; 615; 649; 727; 839; 917

919) 133; 198; 255; 312; 377; 434; 499; 556; 613; 678; 735; 857; 914; 979

920) 312; 323; 334; 345; 356; 367; 378; 389; 814; 825; 836; 847; 858; 869

921) 113; 125; 137; 149; 314; 326; 338; 515; 527; 539; 716; 728; 917; 929

922) 122; 154; 186; 323; 355; 387; 524; 556; 588; 725; 757; 789; 926; 958

923) 133; 187; 212; 266; 345; 399; 424; 478; 557; 636; 715; 769; 848; 927

924) 123; 157; 235; 269; 313; 347; 425; 459; 537; 615; 649; 727; 839; 917

925) 122; 175; 233; 286; 344; 397; 455; 513; 566; 624; 677; 788; 846; 899; 957

926) 154; 222; 276; 344; 398; 412; 466; 534; 588; 656; 724; 778; 846; 914; 968

927) 133; 187; 255; 323; 377; 445; 499; 513; 567; 635; 689; 757; 825; 879; 947

928) 122; 176; 244; 298; 312; 366; 434; 488; 556; 624; 678; 746; 814; 868; 936

929) 211; 232; 253; 274; 295; 512; 533; 554; 575; 596; 813; 834; 855; 876; 897

930) 112; 146; 235; 269; 324; 358; 413; 447; 536; 625; 659; 714; 748; 837; 926

931) 312; 323; 334; 345; 356; 367; 378; 389; 713; 724; 735; 746; 757; 768; 779

932) 213; 224; 235; 246; 257; 268; 279; 515;
526; 537; 548; 559; 817; 828; 839

933) 112; 135; 158; 213; 236; 259; 314; 337;
415; 438; 516; 539; 617; 718; 819

934) 112; 123; 134; 145; 156; 167; 178; 189;
515; 526; 537; 548; 559; 918; 929

935) 153; 211; 264; 322; 375; 433; 486; 544;
597; 655; 713; 766; 824; 877; 935; 988

936) 143; 196; 254; 312; 365; 423; 476; 534;
587; 645; 698; 756; 814; 867; 925; 978

937) 144; 187; 223; 266; 345; 388; 424; 467;
546; 589; 625; 668; 747; 826; 869; 948

938) 123; 157; 212; 246; 335; 369; 424; 458;
513; 547; 636; 725; 759; 814; 848; 937

939) 112; 124; 136; 148; 313; 325; 337; 349;
514; 526; 538; 715; 727; 739; 916; 928

940) 132; 185; 243; 296; 312; 354; 412; 465; 523;
576; 634; 687; 745; 798; 856; 914; 967

941) 122; 165; 244; 287; 323; 366; 445; 488; 524;
567; 646; 689; 725; 768; 847; 926; 969

942) 143; 186; 222; 265; 344; 387; 423; 466; 545;
588; 624; 667; 746; 789; 825; 868; 947

943) 124; 147; 213; 236; 259; 325; 348; 414; 437;
526; 549; 615; 638; 727; 816; 839; 928

944) 134; 168; 223; 257; 312; 346; 435; 469; 524;
558; 613; 647; 736; 825; 859; 914; 948

945) 122; 165; 244; 287; 323; 366; 445; 488; 524;
567; 646; 689; 725; 768; 847; 926; 969

946) 212; 332; 424; 452; 516; 544; 572; 636; 664;
692; 728; 756; 784; 848; 876; 968; 996

947) 212; 223; 234; 245; 256; 267; 278; 289; 514;
525; 536; 547; 558; 569; 816; 827; 838; 849

948) 211; 232; 253; 274; 295; 422; 443; 464; 485;
612; 633; 654; 675; 696; 823; 844; 865; 886

949) 133; 176; 212; 255; 298; 334; 377; 413; 456; 499;
535; 578; 614; 657; 736; 779; 815; 858; 937

950) 112; 135; 158; 224; 247; 313; 336; 359; 425; 448;
514; 537; 626; 649; 715; 738; 827; 916; 939

951) 112; 135; 158; 224; 247; 313; 336; 359; 425; 448;
514; 537; 626; 649; 715; 738; 827; 916; 939

952) 133; 176; 212; 255; 298; 334; 377; 413; 456; 499;
535; 578; 614; 657; 736; 779; 815; 858; 937

953) 143; 186; 211; 254; 297; 322; 365; 433; 476; 544;
587; 612; 655; 698; 723; 766; 834; 877; 945; 988

954) 122; 165; 233; 276; 344; 387; 412; 455; 498; 523;
566; 634; 677; 745; 788; 813; 856; 899; 924; 967

955) 123; 146; 169; 212; 235; 258; 324; 347; 413; 436;
459; 525; 548; 614; 637; 726; 749; 815; 838; 927

956) 212; 223; 234; 245; 256; 267; 278; 289; 513; 524; 535;
546; 557; 568; 579; 814; 825; 836; 847; 858; 869

957) 121; 142; 163; 184; 311; 332; 353; 374; 395; 522; 543;
564; 585; 712; 733; 754; 775; 796; 923; 944; 965; 986

958) 132; 164; 196; 211; 243; 275; 322; 354; 386; 433; 465; 497;
512; 544; 576; 623; 655; 687; 734; 766; 798; 813; 845; 877;
924; 956; 988

959) 122; 154; 186; 233; 265; 297; 312; 344; 376; 423; 455; 487;
534; 566; 598; 613; 645; 677; 724; 756; 788; 835; 867; 899;
914; 946; 978

960) 32; 64; 96

961) 114; 416; 718

962) 214; 327; 617

963) 32; 64; 96

964) 216; 229; 419

965) 114; 229; 317

966) 115; 128; 419

967) 139; 228; 317

968) 225; 259; 416

969) 31; 52; 73; 94

970) 138; 315; 428; 718

971) 211; 243; 275; 533; 565; 597; 823; 855; 887

972) 314; 321; 628; 635; 642; 949; 956; 963

973) 414; 625; 751; 836; 962

974) 522; 614; 961

975) 312; 523; 734; 919; 945; 971

976) 322; 441; 516; 635; 754; 829; 863; 948; 992

977) 111; 332; 425; 518; 553; 646; 681; 739; 774; 867; 995

978) 514; 533; 552; 571; 927; 946; 965; 984

979) 212; 351; 443; 535; 582; 627; 674; 719; 766; 858; 997

980) 361; 452; 543; 634; 725; 816; 995

981) 211; 414; 531; 617; 734; 851; 937

982) 311; 613; 741; 915

983) 513; 632; 751; 916

984) 431; 513; 752; 834; 916; 991

985) 321; 513; 532; 551; 724; 743; 762; 781; 916; 935; 954; 973; 992

986) 221; 341; 461; 514; 581; 634; 754; 874; 927; 994

987) 221; 351; 413; 481; 543; 673; 735; 865; 927; 995

988) 731; 823; 915

989) 722; 914

990) 521; 914

991) 311; 723; 842; 961

992) 211; 532; 614; 724; 853; 916; 982

993) 211; 561; 652; 743; 834; 925

994) 312; 341; 533; 562; 591; 725; 754; 783; 917; 946; 975

995) 714; 823; 932

996) 913

997) 311; 613; 832; 915

998) 722; 914

999) 311; 512; 713; 851; 914

1000)   211; 551; 642; 733; 824; 891; 915; 982

1001)   411; 721; 914

1002)   611; 912

1003)   511; 712; 913

1004)   411; 621; 831; 914

1005)   621; 913

1006)   311; 512; 713; 741; 914; 942

1007)   321; 513; 532; 551; 724; 743; 762; 781; 916; 935; 954; 973; 992

1008)   611; 913

1009)   612; 821; 914

1010)   912

1011)   612; 721; 914

1012)   311; 512; 631; 713; 832; 914; 951

1013)   154; 265; 376; 487; 598; 612; 723; 834; 945

1014)   133; 255; 377; 412; 499; 534; 656; 778; 813; 935

1015)   111; 154; 197; 233; 276; 312; 355; 398; 434; 477;
        513; 556; 599; 635; 678; 714; 757; 836; 915; 958

To show the solutions in the Variety Pack, the entire problem will be shown.


```
1016)   32    Also:   36    1017)   33    1018)   36    1019)   39
      X   9          X   8        X   6        X   3        X   2
       2 8 8         2 8 8        1 9 8        1 0 8         7 8
```

```
1020)   38    1021)   34    1022)   37    1023)   32    1024)   47
      X   6         X   7        X   4        X   4        X   7
       2 2 8         2 3 8        1 4 8        1 2 8        3 2 9
```

```
1025)   68    1026)   77    1027)   84    1028)   46    1029)   74
      X   8         X   4        X   8        X   6        X   9
       5 4 4         3 0 8        6 7 2        2 7 6        6 6 6
```

```
1030)   46    1031)   38    1032)   34    1033)   98    1034)   87
      X   9         X   7        X   9        X   6        X   8
       4 1 4         2 6 6        3 0 6        5 8 8        6 9 6
```

```
1035)   89    1036)   34    1037)   27    1038)   328    1039)  44478
      X   2         +  38        + 5 5        + 5 9 3         + 1 8 5 9 3
       1 7 8         7 2          8 2          9 2 1          6 3 0 7 1
```

```
1040)  7 3 5     1041)  8 0 4     1042)  9 4 2 0 9 2   1043)  8 1 3     1044)  8 0 7
      -- 4 9 6          -- 6 1 5         -- 3 9 6 9 9 4        -- 2 7 9          -- 3 0 8
         2 3 9             1 8 9            5 4 5 0 9 8            5 3 4             4 9 9


1045)  7 4 6     1046)  8 4 1     1047)   2 5    1048)   2 2 9    1049)   5 8 5
      -- 2 4 8          -- 3 8 2          + 6 9           + 4 8 4          + 2 5 6
         4 9 8             4 5 9             9 4             7 1 3             8 4 1


1050)  6 9 8     1051)   6 3     1052)   6 8    1053)   6 8     1054)   6 2
      + 2 3 6           X   8           X   6          X   3           X   9
         9 3 4             5 0 4           4 0 8           2 0 4            5 8 8


1055)   3 5     Also:   7 0
       X  2 4          X  1 2
          1 4 0           1 4 0
           7 0             7 0
          8 4 0           8 4 0
```

With the following solution section, only three solutions will be shown. However, the puzzles themselves may have many more solutions than that. Therefore, you may have discovered a valid solution for a puzzle that is not shown in this answer section.

1056. 521; 714; 832

1057. 211; 532; 853

1058. 235; 459; 618

1059. 723; 842; 961

1060. 513; 632; 751

1061. 771; 844; 982

1062. 224; 336; 448

1063. 412; 531; 925

1064. 523; 818; 945

1065. 314; 527; 919

1066. 111; 355; 746

1067. 522; 614; 961

1068. 624; 835; 971

1069. 324; 659; 817

1070. 411; 822 [Only Two]

1071. 531; 833 [Only Two]

1072. 241; 532; 945

1073. 2113; 3236; 5227

1074. 2134; 5437; 6269

1075. 4348; 5269; 7138

1076. 6114; 8333; 9523

1077. 3231; 6326; 9674

1078. 7132; 8322; 9233

1079. 7141; 8233; 9325

1080. 1157; 1429; 2115

1081. 6213; 7124; 9233

1082. 2141; 3593; 4282

1083. 2891; 7981; 8762

1084. 8171; 9137; 9531

1085. 3234; 8546; 9779

1086. 5317; 7628; 9718

1087. 5197; 8375; 9266

1088. 5311; 8345; 9324

1089. 4114; 6115; 8116

1090. 7313; 8233; 9341

1091. 4113; 8512; 9317

1092. 6266; 7135; 7311

1093. 7249; 8336; 9216

1094. 3139; 5226; 8327

1095. 3121; 6413; 7218

1096. 7364; 8453; 9115

1097. 4256; 6711; 9856

1098. 6769; 7836; 9618

1099. 3112; 8534; 9623

1100. 4129; 7313; 8315

1101. 3179; 8379; 9368

1102. 2326; 3248; 6149

1103. 2324; 3425; 4113

1104. 2215; 2327; 2439

1105. 1762; 3217; 4555

1106. 5311; 8136; 9533

1107. 1213; 3257; 5349

1108. 2124; 2428; 3338

1109. 6395; 8511; 9663

1110. 5125; 8324; 9226

1111. 3432; 5543; 9765

1112. 5384; 8621; 9523

1113. 5255; 8444; 9168

1114. 2428; 4125; 5158

1115. 2267; 3427; 6116

1116. 5114; 8171; 9531

1117. 2113; 2225; 2449

1118. 4351; 6438; 7228

1119. 3123; 4146; 7213

1120. 4121; 7128; 9237

1121. 7132; 8313; 9225

1122. 1735; 5341; 7438

1123. 8113; 8221; 9411

1124. 7314; 8143; 9235

1125. 7145; 7226; 9361

1126. 4585; 6435; 9757

1127. 7233; 8251; 9417

1128. 7232; 8186; 9374

1129. 6266; 7311; 8355

1130. 2427; 4428; 6429

1131. 6112; 8122; 9213

1132. 8211; 9131 [Only Two]

1133. 1179; 3315; 8138

1134. 7229; 8336; 9621

1135. 2335; 4113; 8337

1136. 3437; 5438; 7439

1137. 1213; 2166; 2235

1138. 7231; 8142; 9315

1139. 2113; 3247; 4226

1140. 2225; 4338; 6339

1141. 1429; 2239; 3128

1142. 2154; 4233; 6312

1143. 6211; 8354; 9365

1144. 4112; 7313; 9241

1145. 7139; 8587; 9556

1146. 3125; 6117; 9724

1147. 5311; 8512; 9262

1148. 3127; 6326; 8414

1149. 3237; 4148; 5328

1150. 5217; 6724; 8419

1151. 7233; 8324; 9612

1152. 6131; 8125; 9143

1153. 6114; 8215; 9422

1154. 6343; 7235; 8127

1155. 4254; 5375; 6496

1156. 5186; 8125; 8364

1157. 7231; 8323; 9415

1158. 6112; 8221; 9213

1159. 2112; 6576; 7338

1160. 2328; 3239; 4116

1161. 3114; 8538; 9626

1162. 7322; 8314; 9125

1163. 4247; 5358; 6469

1164. 5354; 7768; 8897

1165. 7243; 8465; 8796

1166. 5131; 8322; 9351

1167. 5526; 7342; 8233

1168. 7241; 8223; 9422

1169. 7233; 8521; 9155

1170. 7221; 8123; 9214

1171. 7124; 8344; 9213

1172. 6258. 7224; 9225

1173. 6255; 8223; 9256

1174. 7141; 9234; 9521

1175. 8122; 8311; 9114

1176. 8321; 9114; 9511

1177. 4111; 8131; 9223

1178. 7981; 8444; 8531

1179. 1835; 2952; 4512

1180. 5463; 7594; 9481

1181. 2353; 2561; 9392

1182. 4268; 8116; 9315

1183. 5211; 8212; 9124

1184. 5453; 7543; 9785

1185. 7299; 8345; 9512

1186. 6214; 9129; 9325

1187. 5356; 6126; 9216

1188. 5458; 7365; 8352

1189. 8171; 9262; 9531

1190. 5132; 7242; 9352

1191. 34339; 43126; 65238

1192. 52211; 72136; 85115

1193. 51199; 82344; 93311

1194. 12127; 22139; 31128

1195. 25136; 32146; 43215

1196. 44381; 53282; 62183

1197. 62178; 82165; 93188

1198. 62176; 81386; 92354

1199. 51297; 83186; 91194

1200. 23362; 45186; 53395

1201. 75346; 83438; 92142

1202. 73125; 82338; 91514

1203. 62187; 81245; 92322

1204. 62127; 73113; 91513

1205. 41314; 75315; 89226

1206. 12113; 43126; 52327

1207. 61527; 73438; 95219

1208. 21149; 82317; 94315

1209. 72116; 81525; 91539

1210. 51131; 81235; 82321

1211. 65324; 86333; 97443

1212. 52214; 73327; 91425

1213. 82565; 83674; 91789

1214. 21953; 65624; 82328

1215. 72133; 82112; 91125

1216. 22234; 34152; 61151

1217. 31113; 53115; 81216

1218. 61348; 72213; 82117

1219. 51325; 63225; 75125

1220. 11316; 31127; 51117

1221. 14142; 33124; 72411

1222. 72127; 81191; 93227

1223. 51214; 51237; 82125

1224. 64951; 75421; 83115

1225. 72284; 83142; 92311

1226. 71166; 82122; 92155

1227. 21122; 63366; 84488

1228. 63188; 83366; 93455

1229. 22316; 43339; 61116

1230. 52234; 52268; 83323

1231. 51211; 71299; 94222

1232. 23315; 33538; 62349

1233. 22271; 23264; 32136

1234. 22226; 41125; 61126

1235. 11155; 23327; 33359

1236. 53111; 71247; 94333

1237. 52367; 82634; 83114

1238. 72257; 83234; 92236

1239. 41134; 72234; 94145

1240. 73568; 85345; 96376

1241. 52311; 75198; 93267

1242. 72112; 81235; 91369

1243. 62237; 84138; 93239

1244. 71113; 81139; 91127

1245. 71336; 82236; 93136

1246. 61445; 62312; 83137

1247. 62135; 74134; 91457

1248. 72212; 83123; 93158

1249. 55133; 64216; 73315

1250. 42133; 71138; 93236

1251. 72847; 82317; 92537

1252. 73411; 82125; 94323

1253. 52211; 81515; 93227

1254. 81149; 81224; 91126

1255. 51221; 72551; 82416

1256. 16191; 49871; 69211

1257. 31255; 46236; 93332

1258. 73544; 83434; 91455

1259. 51348; 72438; 93416

1260. 61227; 73127; 92314

1261. 61113; 71126; 82125

1262. 52116; 72315; 91291

1263. 21116; 75127; 86138

1264. 32212; 31214; 85525

1265. 62121; 71132; 92123

1266. 63464; 82283; 95345

1267. 22355; 42511; 83667

1268. 31112; 33544; 33687

1269. 22383; 33395; 71276

1270. 612226; 731215; 813417

1271. 431145; 854247; 945115

1272. 721352; 832434; 933542

1273. 211186; 753557; 964324

1274. 121158; 122138; 222219

1275. 121411; 234915; 312514

1276. 633229; 641112; 953149

1277. 732413; 834314; 943537

1278. 612431; 711863; 923652

1279. 712314; 812223; 921558

1280. 742519; 861541; 971637

1281. 422415; 725645; 934756

1282. 411617; 721948; 821321

1283. 216322; 523415; 732524

1284. 421625; 832746; 952414

1285. 412441; 853525; 972839

1286. 443234; 466264; 867385

1287. 222278; 332287; 523382

1288. AJ 531285; 834153; 935241*

1289. 711214; 821116; 934328

1290. 321112; 863324; 947939

1291. 132413; 373936; 523616

1292. 712143; 721223; 922314

1293. 411578; 743289; 823564

1294. 185513; 351111; 393252

1295. 311341; 531967; 842765

1296. 234417; 625139; 727249

1297. 315313; 832138; 931113

1298. 621496; 822111; 931285

1299. AJ 421351; 721274; 824232

1300. 118141; 796126; 874214

1301. 232663; 342693; 453991

1302. 2195111; 8115659; 9753446

1303. 7121253; 8111125; 9232121

1304. 7116333; 8318611; 9232111

1305. 7163275; 8521132; 9342145

1306. 5342455; 6254367; 7523152

1307. AJ 2111111; 7438978; 9543287

1308. AJ 4113411; 8325656; 9253267

1309. 2345636; 3456945; 5432555

1310. 5132473; 8233452; 9212464

1311. 2341556; 3542756; 5321488

1312. AJ 1113244; 5327973; 7218662

1313. 3348212; 5729164; 7418224

1314. 52311653; 73452847; 92734945

1315. 67428332; 72539723; 83918948

1316. 35467359; 78373238; 93258435

1317. 43587966; 72495943; 92195885

1318. 74525227; 85439137; 93827329

1319. 34921314; 53847227; 72936125

1320. AJ 14311113; 56473143; 73524393

1321. 3312; 3711; 4141; 4422; 4821; 5133; 5251; 5414; 5813; 5931; 6125; 6243; 6361; 6524; 6642; 6923; 7235; 7353; 7471; 7516; 7634; 7752; 7915; 8182; 8227; 8345; 8463; 8581; 8626; 8744; 8862; 9174; 9219; 9292; 9337; 9455; 9573; 9618; 9691; 9736; 9854; 9972; 5532; 7117

1322. 1121; 1159; 1244; 1367; 1452; 1575; 1698; 1783; 1991; 2145; 2268; 2353; 2476; 2561; 2599; 2684; 2892; 3131; 3169; 3254; 3377; 3462; 3585; 3793; 4155; 4278; 4363; 4486; 4571; 4694; 5141; 5179; 5264; 5387; 5472; 5595; 6165; 6288; 6373; 6496; 6581; 7151; 7189; 7274; 7397; 7482; 8175; 8298; 8383; 8591; 9161; 9199; 9284; 9492 (54 Solutions Total for this last puzzle)

* AJ – A-Jewel's solution. (Read Acknowledgments)